19th Joint Workshop on Electron Cyclotron Emission and Electron Cyclotron Resonance Heating 2016 (EC-19)

EPJ Web of Conferences Volume 147 (2017)

Gandhinagar, India
4 - 7 April 2016

Editors:

B.K. Shukla
H.B. Pandya
S.L. Rao

ISBN: 978-1-5108-4513-8

Printed from e-media with permission by:

Curran Associates, Inc.
57 Morehouse Lane
Red Hook, NY 12571

Some format issues inherent in the e-media version may also appear in this print version.

Printed by Curran Associates, Inc. (2017)

For additional information, please contact EDP Sciences – Web of Conferences at the address below.

EDP Sciences – Web of Conferences
17, Avenue du Hoggar
Parc d'Activité de Courtabœuf
BP 112
F-91944 Les Ulis Cedex A
France

Phone: +33 (0) 1 69 18 75 75
Fax: +33 (0) 1 69 28 84 91

contact-edps@webofconferences.org

Additional copies of this publication are available from:

Curran Associates, Inc.
57 Morehouse Lane
Red Hook, NY 12571 USA
Phone: 845-758-0400
Fax: 845-758-2633
Email: curran@proceedings.com
Web: www.proceedings.com

TABLE OF CONTENTS

EC-19

19th Joint workshop on Electron Cyclotron Emission and Electron Cyclotron Resonance Heating

April 04-07, 2016

IPR, India

List of Participants

1	J. Atul	32	D. Parmar
2	M. Austin	33	J. Patel
3	S. Banerjee	34	H. Patel
4	D. Bora	35	Pavel
5	M. Brookman	36	B. Plaum
6	O. Chellai	37	F. Poli
7	M. Choi	38	S. Pradhan
8	S. Danani	39	N. Rajanbabu
9	G. Denisov	40	S. L. Rao
10	S. Denk	41	V. Rathod
11	S. Dilip	42	K. Sathyanarayana
12	E. Du Toit	43	F. Saul
13	M. Fontana	44	A. Sawant
14	G. Gantenebein	45	R. Shah
15	S. Garavaglia	46	H. Shah
16	N Gupta	47	A. Sharma
17	J. Jelonnek	48	B. K. Shukla
18	J. Jeong	49	V. Siju
19	M. Joung	50	A. Singh
20	P. K. Kaw	51	N. P. Singh
21	N. Khandelval	52	A. Sinha
22	A. Kohn	53	T. Stange
23	S. V. Kulkarni	54	J. Stober
24	R. Kumar	55	D. Strauss
25	R Kumar	56	D. C. Strauß
26	C. H. Lechte	57	P. Subhash
27	J. Lohr	58	E. Tay
28	K. Mishra	59	G. Taylor
29	H. Mistry	60	T. Tsujimura
30	D. Moseev	61	V. Udinstev
31	J. Oosterbeek	62	E. Westerhof

Preface

The 19th Joint workshop on Electron Cyclotron Emission and Electron Cyclotron Resonance Heating (EC-19) was held in Gandhinagar, Gujarat, India on April 04-07, 2016. The workshop was hosted by Institute for Plasma Research (IPR), Bhat, Gandhinagar, India.

The workshop covered four major areas: Electron cyclotron wave (EC-wave) theory, Electron Cyclotron Emission (ECE), Experiments on ECE and ECRH on various tokamaks & Stellarators and Technology associated with ECE and ECRH. There were 12 sessions in four days of the workshop, which included 8 technical sessions, 2 poster sessions, inauguration and summary sessions. The contribution in all sessions are summarised by Dr. M. E. Austin and Dr. Burkhard Plaum. These summaries can be found in the beginning of the relevant section.

The EC-19 workshop at IPR India was a successful workshop with high quality of technical contributions from 63 participants from 12 countries including USA, Russia, Germany, The Netherlands, Japan, Korea, France, Italy and India. The workshop witnessed very insightful discussions during the oral and poster sessions. The main topics that were covered during this workshop included ECRH/ECCD and ECE in fusion plasmas. The technology development in the field of ECRH and ECE was covered in detail. The Gyrotron developments in Russia and Europe were also presented and discussed in detail. The ECRH experiments on various tokamaks including ASDEX, DIII-D & KSTAR were also discussed. The EC systems on various tokamaks were presented by several participants. The other important topics that were discussed include launcher design and other high power microwave systems. The contributions of the workshop received in the form of papers were compiled for publication as a Proceeding for EC-19 workshop. All the papers received from the authors are reviewed by the committee and now published online at EPJ Web Conference of Vol. XX (2017).

The local organizing committee chaired by Prof. Dhiraj Bora did an excellent job in organizing a successful workshop. The participants were offered good facilities with traditional Indian food. The hall arrangements for conducting the sessions at Hotel Narayani, Gandhinagar Gujarat provided the comfortable platform for the workshop. There was a fantabulous cultural program, which highlighted the culture of entire India, this program became memorable for EC-19 participants. I would like to thank my colleagues of International Program committee for their support and finalizing the agenda, program for the workshop, selecting the invited and oral contributions and chairing the various scientific sessions

I would like to thank the sponsors who financially contributed for the workshop: IPR India, Gycom Russia, DRDO India were the key financial supporter for the workshop.

The members of local organising committees were:

Prof. Dhiraj Bora (Chairman)
Braj K. Shukla (Secretary)
R. Jha
Hitesh Pandya
S. Laxmikant Rao

Pramod Kumar Sharma
A. Varadharajulu
Ujjwal Baruh
Chhaya Chavda
P. K. Atrey
Raj Singh
H. K. Sharma
A. V. Ravi Kumar
J. D. Patel

The members of International Program Committee were
Egbert Westerhof (Chairman)
Max Austin (IFS University of Texas at Austin, USA)
Young-Soon Bae (NFRI, Korea)
Dhiraj Bora (IPR, India)
Gregory Denisov (IAP, Nizhny Novgorod, Russia)
Daniela Farina (CNR Milano, Italy)
Franco Gandini (ITER IO)
Mei Huang (SWIP, China)
Shin Kubo (NIFS, Japan)
John Lohr (GA San Diego, USA)
Burkhard Plaum (University Stuttgart)
Braj K. Shukla (IPR, India)
Victor Udintsev (ITER IO)

I would like to thanks the reviewers for reviewing the manuscripts of EC-19 workshop. I would like to thank the team of editors:
Hitesh B. Pandya
S. Laxmikant Rao
Suman Danani
Vipal Rathod
Hardik Mistry
Egbert Westerhof
Braj K. Shukla

Finally once again, I would like to thank my all the national and international colleagues, who helped in successfully organizing the EC-19 workshop in India. My best wishes for the next workshop EC-20 in Greifswald.

Braj K. Shukla
IPR, India, June 2017

edp sciences

EPJ Web of Conferences

Statement of Peer review

In submitting conference proceedings to EPJ Web of Conferences, the editors of the proceedings certify to the Publisher that they adhere to the **Policy on Publishing Integrity** of the journal in order to safeguard good scientific practice in publishing.

1. All articles have been subjected to peer review administered by the proceedings editors.
2. Reviews have been conducted by expert referees, who have been requested to provide unbiased and constructive comments aimed, whenever possible, at improving the work.
3. Proceedings editors have taken all reasonable steps to ensure the quality of the materials they publish and their decision to accept or reject a paper for publication has been based only on the merits of the work and the relevance to the journal.

Title, date and place of the conference

EC-19 (19th Joint workshop on Electron Cyclotron Emission and Electron Cyclotron Resonance Heating)
April 04-07, 2016
IPR, Gandhinagar, Gujarat, India

Proceedings editor(s):

Braj K. Shukla, Hitesh B. Pandya, S.L. Rao.
Egbert Westerhof, Vipal Rathod, Suman Danani &
Hardik Mistry.

Date and editor's signature

(Dr. Braj K. Shukla)
23.06.2017

Perturbing microwave beams by plasma density fluctuations

Alf **Köhn**[1,a], Eberhard **Holzhauer**[2], Jarrod **Leddy**[3], Matthew B. **Thomas**[3], and Roddy G. L. **Vann**[3]

[1] *Max Planck Institute for Plasma Physics, Garching, Germany*
[2] *Institute of Interfacial Process Engineering and Plasma Technology, University of Stuttgart, Stuttgart, Germany*
[3] *York Plasma Institute, Department of Physics, University of York, York, U.K.*

Abstract. The propagation of microwaves across a turbulent plasma density layer is investigated with full-wave simulations. To properly represent a fusion edge-plasma, drift-wave turbulence is considered based on the Hasegawa-Wakatani model. Scattering and broadening of a microwave beam whose amplitude distribution is of Gaussian shape is studied in detail as a function of certain turbulence properties. Parameters leading to the strongest deterioration of the microwave beam are identified and implications for existing experiments are given.

1 Introduction

Electromagnetic waves in the microwave regime are widely used for heating and diagnostic purposes in present fusion experiments based on magnetic confinement. In both cases, the microwaves must propagate across the plasma boundary, a region where substantial density fluctuation levels up to 100 % are known to occur [1]. These fluctuations can disturb a traversing microwave beam, basically changing its beam size and thus potentially spoiling heating efficiencies or leading to ambiguous diagnostics results.

Within this project, the perturbing effect of plasma density fluctuations on a propagating microwave beam is investigated by means of full-wave simulations. The advantage of a full-wave treatment as opposed to geometrical optics techniques is that no restricting assumptions about the size of the turbulent density structures or their amplitude need to be made. The disadvantage is the increased requirement for computational resources which is, however, not a limiting issue with the availability of powerful scientifically focussed computational facilities.

Two full-wave codes are applied, IPF-FDMC [2] and EMIT-3D [3], which are both based on a cold plasma description. Parameter scans are performed in which the properties of the turbulence are varied. To be statistically relevant, the full-wave codes require an ensemble average over many density profiles, all having the same average turbulence properties. Such an ensemble of density profiles was created by a Hasegawa-Wakatani drift-wave turbulence model within the BOUT++ framework [4].

This work is the continuation of previous works in which the influence of single blob-like structures on a traversing microwave beam was investigated [3, 5].

[a] e-mail: alf.koehn@ipp.mpg.de

2 The simulation set-up

In this section, the full-wave codes and the computational grid are briefly described. This includes the technique how the fluctuating electron plasma density profiles are obtained.

2.1 The full-wave codes IPF-FDMC and EMIT-3D

Both full-wave codes solve Maxwell's equations together with an equation for the current density in the plasma de-

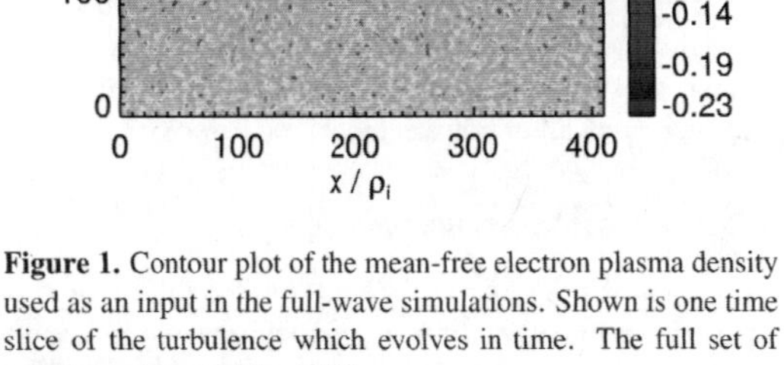

Figure 1. Contour plot of the mean-free electron plasma density used as an input in the full-wave simulations. Shown is one time slice of the turbulence which evolves in time. The full set of turbulence is available at [10].

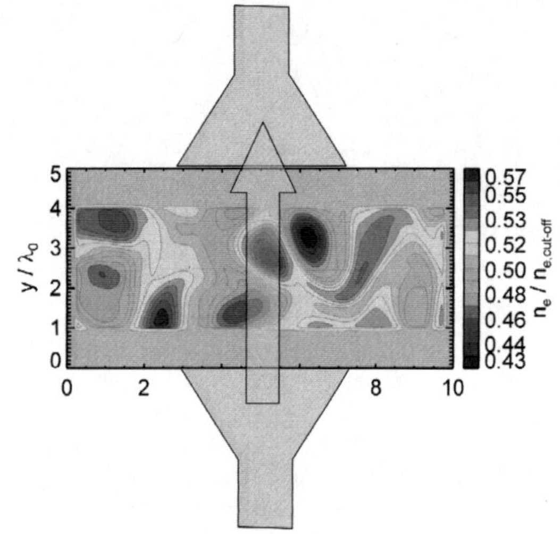

Figure 2. Computational grid used in the full-wave simulations: a Gaussian beam in O-mode polarization is injected at the bottom and detected at the top after interacting with a layer of electron plasma density turbulence (the antenna structures are added for illustration purposes).

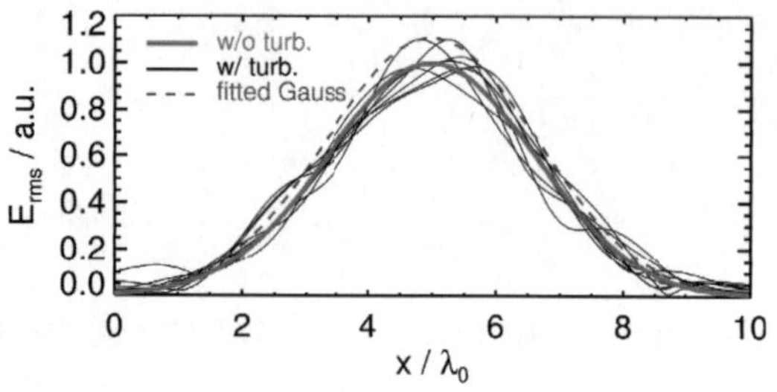

Figure 3. Signal in the detector antenna plane for a few samples belonging to one ensemble for a set of turbulence parameters (black), for the homogeneous case (red), and a Gaussian fitted to all signals from this ensemble (dashed blue).

rived from the fluid equation of motion of the electrons:

$$\frac{\partial}{\partial t}\mathbf{B} = -\nabla \times \mathbf{E} \quad (1)$$

$$\frac{\partial}{\partial t}\mathbf{E} = c_0^2 \nabla \times \mathbf{E} - \mathbf{J}/\epsilon_0 \quad (2)$$

$$\frac{\partial}{\partial t}\mathbf{J} = \epsilon_0 \omega_{pe}^2 \mathbf{E} - \omega_{ce}\mathbf{J} \times \hat{\mathbf{B}}_0 - \nu_e \mathbf{J}, \quad (3)$$

with ω_{pe} the electron plasma frequency, ω_{ce} the electron cyclotron frequency, $\hat{\mathbf{B}}_0$ the unit vector into the direction of the magnetic field, and ν_e the electron collision frequency. The equations are solved with the finite-difference time-domain (FDTD) method on a Cartesian grid (see e.g. Ref. [6] for a comprehensive overview of the FDTD method).

A Gaussian beam in O-mode polarization is injected into the grid using a so-called *soft source*, i.e. the wave electric field is added to the grid at the position of the antenna. The amplitude distribution of the field added reads, neglecting the phase terms, $E(x) = \exp(-x^2/w_0^2)$, where x is the coordinate across the antenna and w_0 the radius of the beam at the waist which is located in the antenna aperture. The antenna itself extends along the whole x-range of the grid to have a smooth field without the unwanted side lobes that would occur at truncated edges. If not mentioned otherwise, a value of $w_0 = 2\lambda_0$ is used with λ_0 the vacuum wavelength of the injected microwave.

2.2 The plasma turbulence

The plasma density turbulence used in the full-wave simulations resembles the type of turbulence which is thought to be the dominant mechanism responsible for the anomalous transport observed in the edge of fusion plasmas, namely drift-wave turbulence [7, 8]. The Hasegawa-Wakatani description [9] is used to model this turbulence within the BOUT++ framework [4]. Figure 1 shows a snapshot of the density turbulence obtained from a BOUT++ run. The data is available for further simulations or benchmarks and can be freely accessed [10].

Since a cold plasma description is used here, the microwave interacts only with the plasma density turbulence which is seen by the microwave as electron density fluctuations. In the time frame of the microwave, the fluctuations are frozen. This is due to the typical frequency scale of the fluctuations which lies in the kHz range (to be compared with the GHz range of the microwaves). In addition, the group velocity of the microwave is orders of magnitudes above the phase velocity of the density structures which can be approximated by the electron diamagnetic drift velocity and reaches values of 10^4 m/s [1]. Therefore, the plasma density fluctuations are taken as fixed non-uniformities of the electron density in the full-wave simulations.

In order to accurately describe the experimental situation, it is not sufficient to perform a single full-wave run with one electron density profile. An ensemble of profiles is required to get statistically relevant results. Such an ensemble was generated with the BOUT++ code as described above [10]. From the spatially large snapshots of the BOUT++ runs, only a small area is needed for a single full-wave run (corresponding to one sample). Within one ensemble, the separate samples correspond each to an area cut from the full grid (the samples do not overlap).

2.3 The computational grid

With IPF-FDMC, simulations are performed on a 2D grid which has a standard size of 10×5 vacuum wavelengths, as indicated in Fig. 2. An emitting antenna is located at the bottom at $y = 0$ and a receiving antenna at the top at $y = 5\lambda_0$. The grid is surrounded by non-radiating boundaries. As can be seen in Fig. 2, the grid has a homogeneous background electron density $n_{e,0}$ with a value of half of the O-mode cut-off density $n_{e,\text{cut-off}}$. The fluctuations are actually located in a layer with a width of $3\lambda_0$, where a smooth transition from the homogeneous background to the turbu-

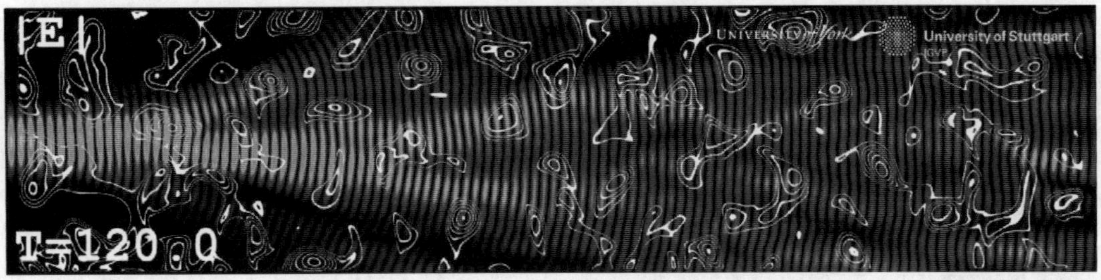

Figure 4. Snapshot of the absolute value of the wave electric of an electromagnetic wave propagating across a plasma with electron density fluctuations, indicated by the white contour lines representing positive perturbations as compared to the background density. The snapshot is taken from a video published at [12].

lence layer is employed over a few grid points to avoid spurious reflections.

The background magnetic field has a normalized value of $Y = \omega_{ce}/\omega_0 = 0.5$ and is directed perpendicular to the grid. Due to the drift-wave nature of the density structures, they are elongated along the magnetic field lines. This resembles the 2D grid used in IPF-FDMC, which assumes no variation in the third dimension. In EMIT-3D, however, full 3D runs are performed with the same grid size in the perpendicular direction (perpendicular to the background magnetic field). In the third dimension, the density structures are assumed to not vary and are thus simply extended along the magnetic field. This allows to compare the 2D simulations of IPF-FDMC with the 3D simulations of EMIT-3D and therefore elaborate if the scattering in the geometry used is basically 2D or 3D in nature. Due to the increased demand in computational resources when going from a 2D simulation to a 3D simulation, the comparison is only carried out for a few dedicated cases. In addition, a few 3D simulations are performed to investigate the effect of an oblique injection of the microwave beam onto the electron density filamentary structures.

3 Data analysis

The full-wave simulations are based on a time-dependent scheme and start with the injection of the microwave beam at the bottom sending antenna. They are stopped when the microwave has propagated across the turbulence layer and a steady state solution is achieved. During the simulations, the wave electric field is continuously recorded at the receiving antenna, which spans across the whole simulation grid, see Fig. 2. More precisely, a time-averaged field is recorded, defined as follows:

$$\tilde{E}_{\text{rms}} = \sum_t \frac{\sqrt{\tilde{E}_x^2 + \tilde{E}_y^2 + \tilde{E}_z^2}}{\sqrt{T}}, \tag{4}$$

with t the time step in the full-wave simulations, T the number of the wave periods passed since the start of the simulation and the superscript $\tilde{}$ referring to the perturbations due to the electron density fluctuations. Figure 3 shows detector antenna signals according to Eq. (4) for a few samples from one ensemble of density profiles. The corresponding signal for the homogeneous case, i.e. without fluctuations is also shown, clearly illustrating the perturbing effect of the fluctuations.

The scattering observed for each sample of density turbulence is quantified by summing up the squared deviations of the $\tilde{E}_{\text{rms}}$ signal to the homogeneous case. Thus, a scattering parameter α is defined:

$$\alpha = \frac{\sum_x \left(\tilde{E}_{\text{rms}} - E_{\text{rms}}\right)^2}{\sum_x E_{\text{rms}}^2}. \tag{5}$$

A separate value of α is obtained for each sample and ensemble averaging is then performed to get the scattering for one set of turbulence parameters.

A second analysis is applied which consists in determining the average detector antenna signal by calculating the arithmetic mean of the $\tilde{E}_{\text{rms}}$ signals from all samples belonging to one ensemble. A Gaussian function is then fitted to the averaged signal, allowing to get the beam size of the averaged perturbed beam, w_{turb}. Thus an average broadening of the microwave beam can be determined. This is important for the case of a localized current drive by injected microwaves with the idea of stabilizing neoclassical tearing modes (NTMs) which can lead to a sudden loss of the confinement in a tokamak [11].

4 Simulation results

Before starting to analyse the scattering described in the previous Section, an illustrative example is presented first. In contrast to the grid shown in Fig. 2, a much larger computational grid is used in order to emphasize the effect of electron density perturbations on a microwave beam. The background density corresponds again to half of the O-mode cut-off density. As shown in Fig. 4, the scattering of the microwave can be clearly seen in this example, leading not only to a deviation of the original, straight beam path but also to a splitting into multiple beams. The simulation was obtained with the 2D full-wave code IPF-FDMC and can be accessed as a video, showing the microwave beam propagating across the grid, at Ref. [12].

For the quantitative analysis of the scattering process, ensemble averaging is required as explained in Sec. 3. To

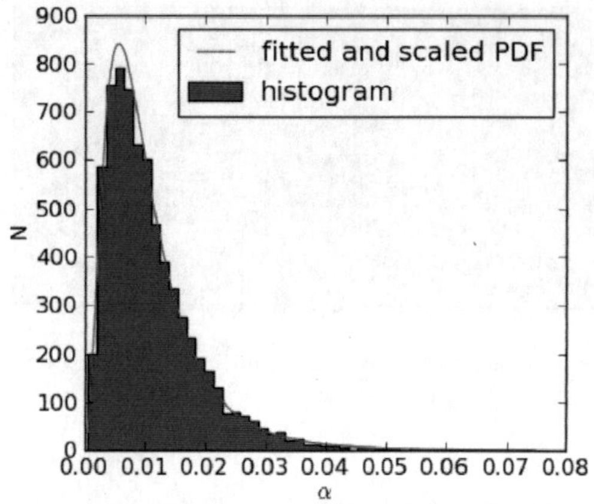

Figure 5. Histogram and fitted PDF of the scattering parameter α for a size of the turbulent density structures of $L_c \approx 0.5\,\lambda_0$. The histogram follows a log-normal distribution.

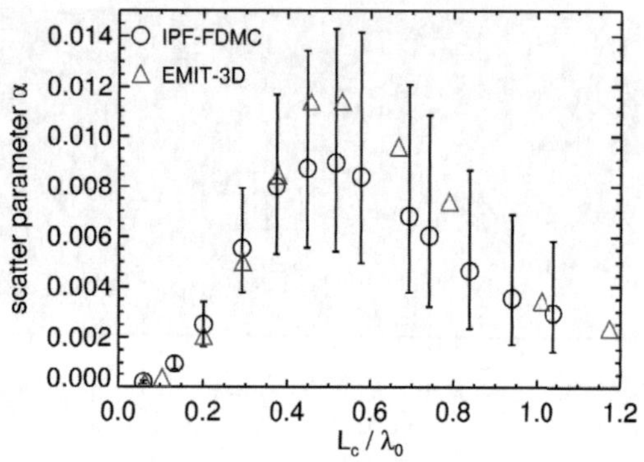

Figure 6. Median of the scatter parameter α, see Eq. (5), as a function of the electron density structure size for an average fluctuation strength of $\sigma \approx 4\,\%$. The error bars correspond to the interquartile range.

ensure that the ensemble is large enough, the average position of the maximum value of the $\tilde{E}_{\mathrm{rms}}$ signal in the detector antenna plane is compared with the position of the maximum for the homogeneous case. The difference should asymptotically approach zero (within error bars). It has been assured that this is the case for all ensembles considered in this paper. The resulting size of an ensemble can be as large as $N = 25,000$. It varies when different average spatial sizes of the electron density structures are considered, where the spatial size is measured as the perpendicular correlation length L_c which is varied from $L_c \approx 0.06\,\lambda_0 \ldots 1.2\,\lambda_0$. With decreasing values of L_c, more density structures fit into the turbulence layer and the spatial average of one sample becomes better reducing the required ensemble size.

According to Eq. (5), $\alpha \geq 0$ is always fulfilled. It is therefore not expected that α follows a normal distribution. Figure 5 shows the obtained distribution for a structure size of $L_c \approx 0.5\,\lambda_0$: it actually follows a log-normal distribution. This applies to all values of L_c and to statistically describe the scatter parameter α, its median and the interquartile range will therefore be used.

Figure 6 shows the scattering as a function of the structure size for an average fluctuation strength of $\sigma \approx 4\,\%$, which is measured as

$$\sigma = \sqrt{\frac{1}{N_{x,y}} \sum_{x,y} (\tilde{n}_e(x,y) - n_0)^2}, \tag{6}$$

with $N_{x,y}$ the number of grid points in the turbulence slice. Looking first at the results from IPF-FDMC, the scattering α exhibits a maximum at around $L_c \approx 0.5\,\lambda_0$. Very large structures exceeding the width of the turbulence layer act effectively as a phase plate and the scattering is therefore expected to approach an asymptotic value as observed. If the density structures are too small they hardly have an effect on the microwave beam (obstacles with a size below $\lambda_0/10$ can generally be considered as not perturbing the microwave). The errorbars correspond to the interquartile range and obviously α has the largest spread where the maximum is located.

The results from EMIT-3D are also shown in the plot and very good agreement is found. Note that the ensemble size is reduced by approximately two orders of magnitude as compared to IPF-FDMC due to the additional third dimension which leads to increased computational time. The agreement shows that the scattering in the geometry used, is basically a 2D process.

The average beam broadening obtained from the same simulations is shown in Figure 7 as a function of L_c. It follows a similar behaviour as α, although the maximum is more pronounced. A broadening of approximately 2 % is found at maximum.

Another important parameter influencing the scattering is the strength of the fluctuation amplitude. A number of parameter scans have been performed with the 2D full-wave code in which the fluctuation strength was varied in a range of $\sigma = 2 \ldots 12\,\%$. The average structure size was $L_c \approx 0.5\,\lambda_0$ in all cases, corresponding to the strongest scattering found in the previous parameter scans, see Fig. 6. The width of the turbulence layer was kept constant. It is found that both α and the average beam broadening as a function of the fluctuation strength σ follow a quadratic increase.

In an additional scan, the width of the turbulence layer is varied in a range from $2 \ldots 7\,\lambda_0$. A linear influence of the width is found on both the scattering parameter α and the average beam broadening.

Further parameter scans performed in the same geometry are described in detail in Ref. [13].

5 Application of simulation results

Microwaves can be used to drive localized toroidal net currents in order to stabilize NTMs [11], as briefly mentioned earlier. The results obtained with the full-wave simulations are now applied to the case of the ASDEX Upgrade tokamak [14] for such a scenario. The microwave

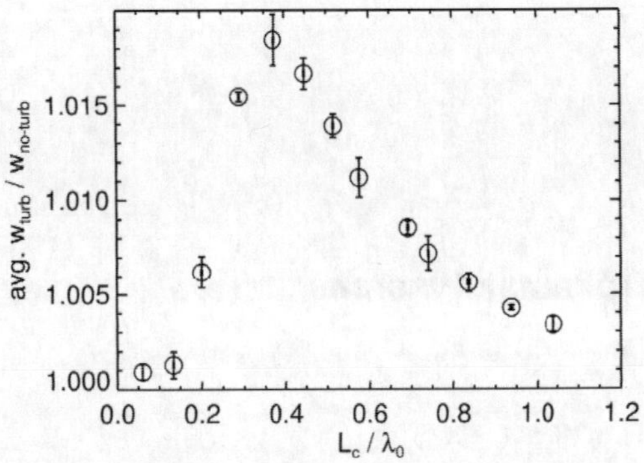

Figure 7. Average beam broadening as a function of the density structure size for an average fluctuation strength of $\sigma \approx 4\,\%$.

frequency is 140 GHz, corresponding to a vacuum wavelength of $\lambda_0 \approx 2$ mm. The turbulence layer is located at the scrape-off layer with a density of approximately 10 % of the cut-off density of the injected microwave [15] and an average size of the perturbing density structures of $L_c \approx 4\,\lambda_0$ [16]. Although the average fluctuation level is with $\sigma \approx 15\,\%$ [17] relatively large, only small perturbations of the microwave are expected due to the large density structure size with respect to the vacuum wavelength and the relatively low background plasma density.

6 Summary

Full-wave simulations of a Gaussian beam in O-mode polarization injected onto a layer of electron plasma density turbulence have been performed. It was shown that the scattering is basically 2D in nature when the microwave beam is injected perpendicular onto the plasma density structures which are elongated along the magnetic field lines, resembling drift-wave turbulence structures. The strongest deterioration of the microwave was found for an average density structure size corresponding to half of the vacuum wavelength of the injected microwave. A square dependence on the strength of the fluctuation and a linear dependence on the width of the turbulence layer was found in the parameter range used here. The simulation results were applied to one example, NTM stabilization in the ASDEX Upgrade tokamak, with the result that the deterioration of the microwave seems not to be important in this case.

Acknowledgements

One of the authors (A. K.) wants to thank Burkhard Plaum for providing him with a useful png-plotting library.

Part of the simulations were performed on the HELIOS supercomputer system at Computational Simulation Centre of International Fusion Energy Research Centre (IFERC-CSC), Aomori, Japan, under the Broader Approach collaboration between Euratom and Japan, implemented by Fusion for Energy and JAEA.

One of the authors (M.B. T.) was funded by the EPSRC Centre for Doctoral Training in Science and Technology of Fusion Energy grant EP/L01663X.

References

[1] S.J. Zweben *et al.*, Plasma Phys. Control. Fusion **49**, S1 (2007). http://dx.doi.org/10.1088/0741-3335/49/7/S01

[2] A. Köhn *et al.*, Plasma Phys. Control. Fusion **50**, 085018 (2008). http://dx.doi.org/10.1088/0741-3335/50/8/085018

[3] T.R.N. Williams *et al.*, Plasma Phys. Control. Fusion **56**, 075010 (2014). http://dx.doi.org/10.1088/0741-3335/56/7/075010

[4] B. Dudson *et al.*, Comp. Phys. Comm. **180**, 1467 (2009). http://dx.doi.org/10.1016/j.cpc.2009.03.008

[5] A. Köhn *et al.*, EPJ Web of Conferences **87**, 01003 (2015). http://dx.doi.org/10.1051/epjconf/20158701003

[6] A. Taflove and S.C. Hagness, *Computational Electrodynamics: the Finite-Difference Time-Domain Method* (Artech House Publishers, Boston, 2000)

[7] Horton W, Phys. Reports **192**, 1 (1990). http://dx.doi.org/10.1016/0370-1573(90)90148-U

[8] Wootton A J *et al.*, Phys. Fluids B **2**, 2879 (1990). http://dx.doi.org/10.1063/1.859358

[9] Wakatani M *et al.*, Phys. Fluids **27**, 611 (1984). http://dx.doi.org/10.1063/1.864660

[10] J. Leddy *et al.*, Plasma density turbulence obtained from a Hasegawa-Wakatani drift-wave turbulence model within the BOUT++ framework. Zenodo. (2016). http://dx.doi.org/10.5281/zenodo.47206

[11] Poli E *et al.*, Nucl. Fusion **55** 013023 (2015). http://dx.doi.org/10.1088/0029-5515/55/1/013023

[12] A. Köhn *et al.*, FDTD simulation of a microwave beam propagating across a plasma with density fluctuations, figshare (2016). https://dx.doi.org/10.6084/m9.figshare.3201115.v1

[13] A. Köhn *et al.*, arXiv (2016). http://arxiv.org/abs/1604.00344

[14] H. Zohm *et al.*, Nucl. Fusion **47** 228 (2007). http://dx.doi.org/10.1088/0029-5515/47/3/010

[15] J. Stober *et al.*, Plasma Phys. Control. Fusion **42** A211 (2000). http://dx.doi.org/10.1088/0741-3335/42/5A/324

[16] G. Fuchert *et al.*, Plasma Phys. Control. Fusion **56** 125001 (2014). http://dx.doi.org/10.1088/0741-3335/56/125001

[17] P. Hennequin *et al.*, 42nd EPS Conference (2014).

DOI: 10.1051/epjconf/201714701002

A Kinetic Study of Microwave Start-up of Tokamak Plasmas

E.J. **du Toit**[1,2,a], M.R. **O'Brien**[2], and R.G.L. **Vann**[1]

[1] *York Plasma Institute, Department of Physics, University of York, York, YO10 5DD, UK*
[2] *Culham Centre for Fusion Energy, Abingdon, OX14 3DB, UK*

Abstract. A kinetic model for studying the time evolution of the distribution function for microwave start-up is presented. The model for the distribution function is two dimensional in momentum space, but, for simplicity and rapid calculations, has no spatial dependence. Experiments on the Mega Amp Spherical Tokamak have shown that the plasma current is carried mainly by electrons with energies greater than 70 keV, and effects thought to be important in these experiments are included, i.e. particle sources, orbital losses, the loop voltage and microwave heating, with suitable volume averaging where necessary to give terms independent of spatial dimensions. The model predicts current carried by electrons with the same energies as inferred from the experiments, though the current drive efficiency is smaller.

1 Introduction

Non-inductive plasma current start-up is a very important area of research for the spherical tokamak (ST) due to a lack of space for a shielded inboard solenoid. Reliable models are needed to predict performance and start-up requirements for present and future STs. This paper describes development of such a model for electron Bernstein waves, and first comparison of its predictions with experimental results.

Electron Bernstein wave (EBW) start-up using a 28 GHz gyrotron capable of delivering 100 kW for up to 0.5 s was demonstrated on the Mega Amp Spherical Tokamak (MAST) [1, 2]. The EBW start-up method employed here relied on the production of low-density plasma by RF pre-ionization around the fundamental electron cyclotron resonance (ECR). The toroidal magnetic field generated by the central rod (CR), generated by a current of about 2 MA, had a value of $B_\phi = 1$ T at a radial position of about 0.4 m. A double mode conversion (MC) was used for EBW excitation. The scheme consisted of MC of the ordinary (O) mode, incident from the low field side of the tokamak, into the extraordinary (X) mode with the help of a grooved mirror-polarizer incorporated in a graphite tile on the CR. The launched Gaussian beam was tilted to the midplane at 10° and hit the CR at the midplane, as illustrated in figure 1. The X-mode reflected from the CR propagated back to the plasma, passed through the ECR and experienced a subsequent slow X to EBW MC near the upper hybrid resonance (UHR). The excited EBW mode was totally absorbed before it reached the ECR, due to the Doppler shifted resonance. Modelling showed that only a small fraction of the injected power (~ 2%) was absorbed as the O- and X-modes, while the main part was converted into and then absorbed as the EBW mode [2].

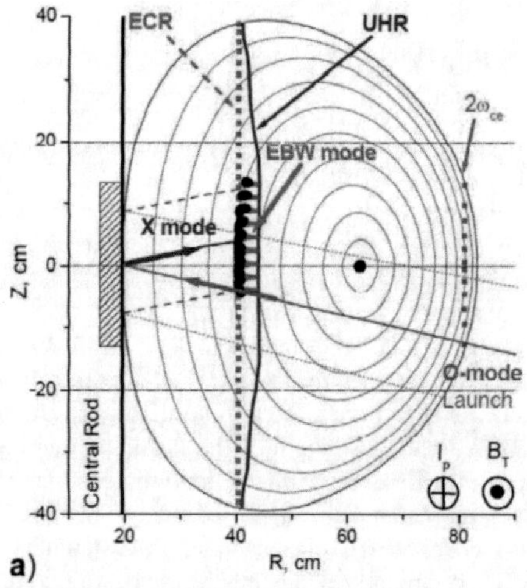

Figure 1. Schematic of the EBW assisted plasma current start-up. The black area at the end of the EBW rays indicates the power deposition zone [1].

The various experiments conducted on EBW assisted start-up are well documented in [1] and [2], as well as the references therein. The main conclusions were: firstly, the formation of closed flux surfaces (CFS) is governed by EBW current drive (CD); secondly, the plasma current is carried predominantly by supra thermal electrons with energies above 70 keV; and thirdly, there exists a linear relationship between the injected RF power and the gen-

[a]e-mail: ejdt500@york.ac.uk

erated plasma current, such that an overall current drive efficiency of 1 A/W was achieved.

Models are important to interpret and predict start-up current drive in STs, and confirm conclusions drawn from experiments. To this end, a start-up model has been developed to study the time evolution of the electron distribution function. For simplicity, and to allow for rapid calculations, the distribution function has zero spatial dependence, while it is two dimensional in momentum space. The effects thought to be important in capturing the main physics during the early stages of the plasma discharge have been included, i.e. particle sources, orbital losses, the loop voltage and EBW heating, with suitable volume averaging where necessary to give terms independent of spatial dimensions.

The paper is structured as follow: an introduction to the kinetic model describing the time evolution of the electron distribution function for studying EBW start-up is presented, with special attention being given to particle losses and EBW heating, and how the spatial dependences are approximated in 0D in these terms. This is followed by a discussion of results and its comparison to experiments done on MAST, and a brief overview of future work.

2 0D2V kinetic start-up model

In general, the electron distribution function depends on space, momentum, and time, $f(\vec{r}, \vec{p}, t)$. The distribution function can be used to calculate the current density,

$$J\left(\vec{r}, \vec{t}\right) = e \int \mathrm{d}^3 p \, v_\parallel \, f(\vec{r}, \vec{p}, t)$$

and the electron density,

$$n_e = \int \mathrm{d}^3 p \, f(\vec{r}, \vec{p}, t).$$

In order to study the time evolution of the distribution function, simplifications are made to ensure the model is tractable and computationally manageable. For this reason, the model is zero dimensional (0D) in space, but spatial dependences for each term are taken into account when calculating the 0D approximations. Further, the time it takes for an electron to complete a gyro-orbit is fast compared to all other timescales, such that the momentum dependence can be captured in two dimensions (2V). The time evolution of the electron distribution function is then studied under several effects,

$$\frac{\partial f}{\partial t} = \text{source} + \text{loss} + \text{EBW heating} + \text{loop voltage} \quad (1)$$

where $f = f(p_\parallel, p_\perp, t)$.

Experiments suggest that the plasma current is carried by supra thermal electrons with energies in the range 70 – 100 keV. Under typical MAST parameters, these electrons collide very infrequently (for electrons with energies above 20 keV, the collision times are longer than 100 ms), and therefore collisions are neglected in this paper, while the terms expected to be most important in studying the time evolution of the distribution function are retained.

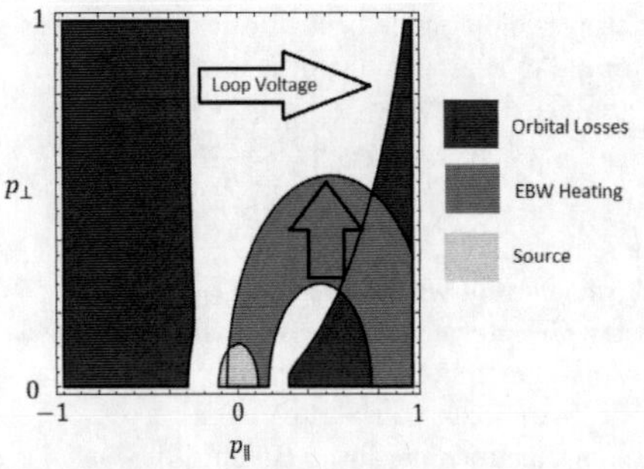

Figure 2. Schematic of the effects under which the time evolution of the distribution function is studied.

Each term acts in a different region of momentum space, as illustrated in figure 2. The source term is due to cold electrons entering the system, resulting mainly from ionization. These electrons are taken to be isotropic in momentum, such that

$$\left(\frac{\partial f}{\partial t}\right)_{\text{source}} = \frac{S_0}{\pi^{3/2} p_0^3} \exp\left[-\frac{p_\parallel^2 + p_\perp^2}{p_0^2}\right]$$

where p_0 is the characteristic momentum of these cold electrons, and S_0 is the rate of change in electron density.

The loss term is due to electrons being lost out of the plasma volume, described in section 2.1. The EBW heating term describes the interaction between the plasma and the injected RF beam, described in section 2.2.

The electric field created by the possible use of the central solenoid results in an acceleration of electrons to higher parallel momentum, and is described by the loop voltage term, given by

$$\left(\frac{\partial f}{\partial t}\right)_{\text{loop voltage}} = -e \frac{V_L}{2\pi R} \frac{\partial f}{\partial p_\parallel}$$

where V_L is the magnitude of the loop voltage, and R is the major radius.

2.1 Orbital Losses

An important part of the start-up scenario is the transition from an open field line configuration to closed flux surfaces (CFS). During the initial start-up phase, the magnetic field lines are considered to be open, and electrons can freely stream along these field lines out of the plasma volume into the vessel walls. This loss is due to the parallel velocity of the electrons, and depends on the vertical strength of the magnetic field, such that a characteristic confinement time for an electron with parallel velocity $v_\parallel$ can be modelled as [3],

$$\tau_\parallel = a_\perp \frac{B_\phi}{B_Z} / v_\parallel$$

where $a_\perp$ is the perpendicular distance to the vessel walls. This term has to be modified to allow for the formation of

CFS, as particle confinement drastically improves when CFS form. An additional factor is therefore added,

$$\tau_{\parallel} = a_{\perp} \exp\left(\frac{I_p}{I_{\text{ref}}}\right)\frac{B_\phi}{B_Z}/v_{\parallel} \tag{2}$$

where I_p is the plasma current and I_{ref} is the magnitude of the plasma current where CFS form [3].

After CFS form, the loss mechanism changes from mainly parallel transport to mainly perpendicular transport. Collisions can transport electrons across CFS, such that some electrons are lost. Bohm diffusion is used to describe perpendicular transport [4], such that a characteristic perpendicular loss time is given by,

$$\tau_{\perp} = \frac{8a^2}{T_e(eV)}B_\phi \tag{3}$$

where a is the minor radius of the plasma and T_e is the energy of the electron.

The confinement time of an electron is therefore given by,

$$\frac{1}{\tau_e} = \frac{1}{\tau_{\parallel}} + \frac{1}{\tau_{\perp}} \tag{4}$$

with the result for an electron of a specific energy shown in figure 3.

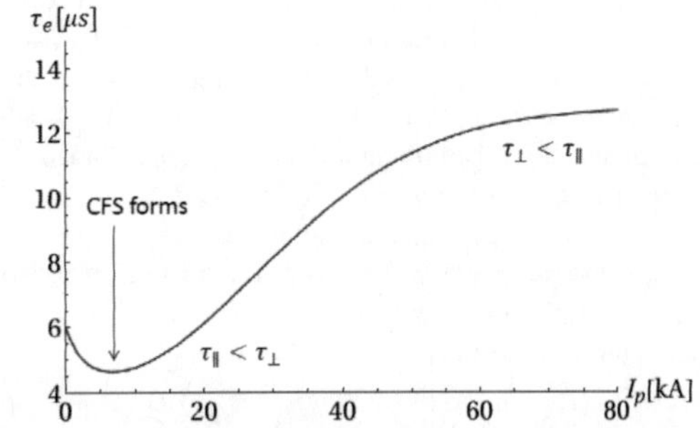

Figure 3. Characteristic particle loss time for an energetic electron (E_k = 50 keV) during start-up phase as a function of plasma current. At small currents, before closed flux surfaces form, parallel transport dominates, while perpendicular transport dominates for large currents.

This loss time, however, does not take into account the magnetic field line structure leading to effects such as trapped electrons. It can therefore only be used as an estimation of the particle loss time, as a function of plasma current, while the momentum dependence has to be added explicitly. For this reason, the distribution function is subjected to a loss term of the form,

$$\left(\frac{\partial f}{\partial t}\right)_{\text{loss}} = -\frac{f}{\tau_e}A(p_{\parallel}, p_{\perp}) \tag{5}$$

where $A(p_{\parallel}, p_{\perp})$ is the functional form of the loss term, containing the momentum dependence, while the loss time τ_e determines the magnitude of the loss term.

The momentum dependence of the loss term is found by studying single particle orbits during an open field line structure. By studying the guiding centre orbits of electrons originating from the same point in space, the time it takes for these electrons to be lost or to complete a confined orbit can be calculated, and plotted as a function of momentum. The time it takes for electrons to be lost out of the plasma volume, calculated by the guiding centre approach, is consistent with the time found from equation (4). Figure 4 shows the initial values of momentum for which electrons complete a confined orbit, while all other electrons will be lost out of the plasma volume. An approximate form for the loss term can then be determined,

$$A(p_{\parallel}, p_{\perp}) = \frac{p_{\parallel}^2 + p_{\perp}^2}{p_{\text{loss}}^2}\begin{cases}\left(1 - \exp\left[-\frac{5p_{\parallel}^2}{4p_{\perp}^2}\right]\right) & p_{\parallel} < 0\\ \left(1 - \exp\left[-\frac{4p_{\parallel}^2}{5p_{\perp}^2}\right]\right) & p_{\parallel} \geq 0\end{cases} \tag{6}$$

where p_{loss} is the characteristic momentum of an electron with loss time τ_e.

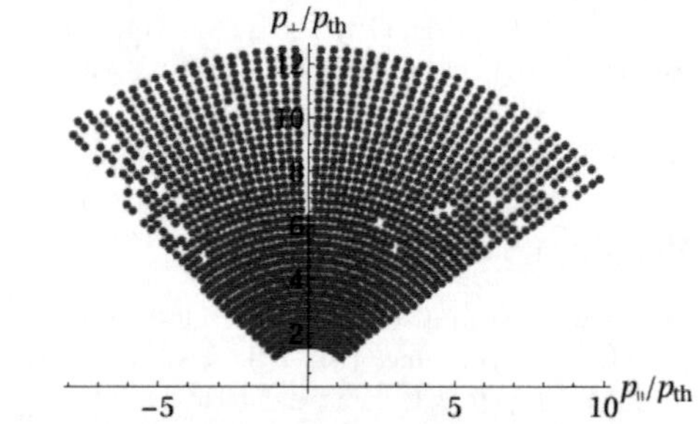

Figure 4. Confinement map for electrons originating from the same point on the poloidal plane show a slight asymmetry. Electrons initially moving against the magnetic field are lost more easily than electrons moving along the magnetic field.

The typical confinement map for electrons show that the confinement time decreases for increasing parallel momentum, while confinement improves when increasing the perpendicular momentum. There is also an asymmetry in the loss term, for electrons moving with and against the magnetic field, which will disappear once CFS form.

2.2 EBW Heating

The interaction between the injected RF beam and the plasma is described by the EBW heating term. This form of heating mainly increases the perpendicular momentum of electrons,

$$\left(\frac{\partial f}{\partial t}\right)_{\text{EBW heating}} = \frac{1}{p_{\perp}}\frac{\partial}{\partial p_{\perp}}p_{\perp} D(p_{\parallel}, p_{\perp})\frac{\partial f}{\partial p_{\perp}} \tag{7}$$

where the diffusion coefficient is assumed to take the form,

$$D(p_{\parallel}, p_{\perp}) = D_0\left\langle \exp\left[-\left(\frac{p_{\parallel} - p_{\parallel 0}}{\Delta p_{\parallel}}\right)^2\right]\right\rangle_{\text{volume average}} \tag{8}$$

and D_0 is a constant to be determined. It is assumed that the EBW heating is localised in momentum space, and that it occurs where the resonance condition ($\omega - k_{\parallel}v_{\parallel} - n\omega_c = 0$) is satisfied. For a particular value of $p_{\perp}$, $p_{\parallel 0}$ is the parallel momentum that satisfies the resonance condition. The absorption width $\Delta p_{\parallel}$ is due to a spread in $k_{\parallel}$, the parallel component of the wave number of the EBW.

EBW heating is not uniform in space, due to the dependence of the relativistic cyclotron frequency $\omega_c = q_e B/m_e\gamma$ on the magnetic field. The spatial dependence of the plasma wave interaction has to be resolved, and this is done by integrating over the plasma in taking a volume average,

$$\left\langle \exp\left[-\left(\frac{p_{\parallel} - p_{\parallel 0}}{\Delta p_{\parallel}}\right)^2\right]\right\rangle_{\text{volume average}} = \frac{1}{V}\int_{\text{plasma}} \mathrm{d}V \exp\left[-\left(\frac{p_{\parallel} - p_{\parallel 0}}{\Delta p_{\parallel}}\right)^2\right]$$

such that the plasma wave interaction is limited to the region in space where the plasma exists. This produces a momentum dependent diffusion term $D(p_{\parallel}, p_{\perp})$ independent of space, illustrated in figure 5. The diffusion term is largest along the resonance curve (the solution to the resonance condition) where absorption will be maximal, but smeared out in momentum space due to the spatial dependence of the resonance condition, as the resonance curve is slightly different for different values of the magnetic field.

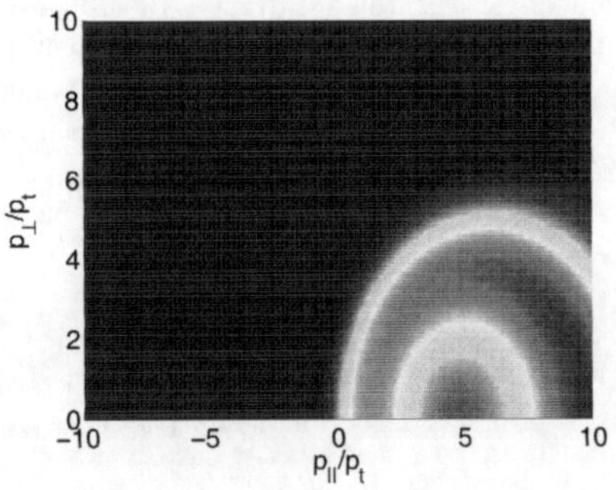

Figure 5. Typical resonance curve for EBW heating shows the location where the plasma wave interaction is prominent. The diffusion term is smeared out in momentum space due to the spatial dependence of the resonance condition, while a maximum value of $p_{\perp}$ beyond which the resonance condition can no longer be satisfied is clearly visible.

The power absorbed by the plasma from the EBW can be calculated from the distribution function,

$$P_d = \frac{1}{2}m_e \int \mathrm{d}V \int \mathrm{d}^3 p \left(\frac{\partial f}{\partial t}\right)_{\text{EBW heating}} \tag{9}$$

where the volume integral is introduced to obtain the total power absorbed, rather than the power density. This can be related to the absorption efficiency of the plasma through the optical depth σ:

$$\sigma P_0 = P_d \tag{10}$$

where P_0 is the injected RF power. In general, radiation can be reflected in the tokamak to give multi-pass absorption, but as the plasma volume is assumed to be small compared to the vessel volume during start-up, this effect is assumed to be small, and absorption is approximated as single pass.

For ECRH, standard expressions for σ exist in the literature (see, for example [5]), such that the absorption efficiency σP_0 can readily be calculated. For EBW, absorption is expected to be much stronger, such that σ would be approximately 1. The value of σ, in this case, would then be related to the amount of power converted to EBW, such that the expression (10) remains the same. The value of D_0 can then be chosen in order to satisfy this equation, by ensuring the distribution function produces the correct value of P_d.

This method can be used to study both ECRH and EBW start-up. The wave parameters in these two cases of heating are different, but the diffusion coefficient is related to the resonance condition, which depends on the particular wave parameters, while the absorption is related to the optical depth in the case of ECRH, or to the amount of power converted to EBW. This is a rather versatile method in that sense, as the terms and their forms do not change depending on the type of heating, but only the values.

3 EBW Start-up

The experiments on MAST summarised above relied on the injection of 100 kW of RF power. Modelling showed that the majority of this power (~ 95%) was absorbed as EBW, which generate large values of $N_{\perp}$ ($N = kc/\omega$), but can have values of $N_{\parallel} \approx 1$ [1]. The electron density during the start-up phase of these experiments ranged from very small values to densities of 10^{18} m^{-3}, and it is expected that the absorption will increase with the electron density until all power is absorbed, i.e. $\sigma \approx 1$.

The simulation iterates over the source term, S_0 (fig. 6) to reproduce experimentally observed densities (fig. 7). At low densities the electron cyclotron resonance (ECR) and upper hybrid resonance (UHR) are close to each other, such that a small amount of power will be converted to EBW. As the density increases, however, the distance between the ECR and UHR increases and more power will be converted to EBW. The power converted to EBW is therefore assumed to be proportional to the electron density, while $\sigma = 1$. The absorbed power is shown in figure 8.

The value of the diffusion coefficient D_0 can be determined such that the power absorbed calculated from the distribution function, P_d, equals the power absorbed found from the optical depth, σP_0. The value of D_0, as a function of time, is shown in figure 9. A theoretical value for D_0 can be calculated, and such a value is a good initial value for D_0, but as the distribution function evolves in time and becomes distorted, the value of D_0 has to be updated to ensure the correct power is absorbed.

It is found that the distribution function flattens off, and therefore the value of D_0 keeps increasing to maintain full power absorption. At low power, the value of D_0 re-

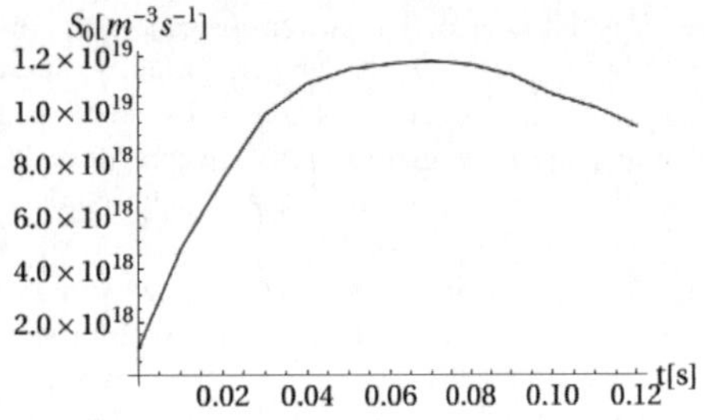

Figure 6. Simulated magnitude of the source term, S_0, which reproduces experimental densities.

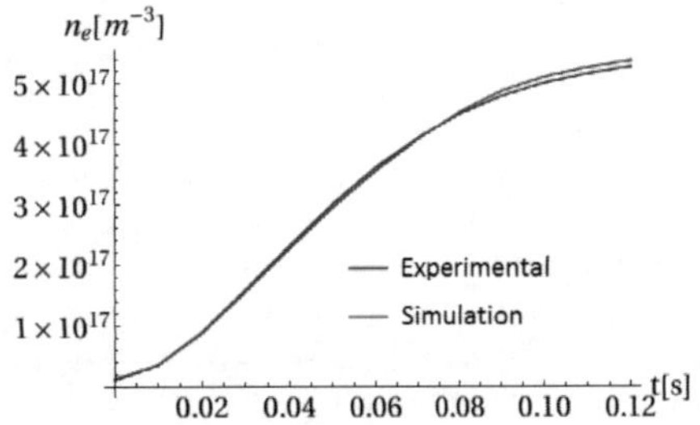

Figure 7. Experimental and simulated electron densities for EBW start-up.

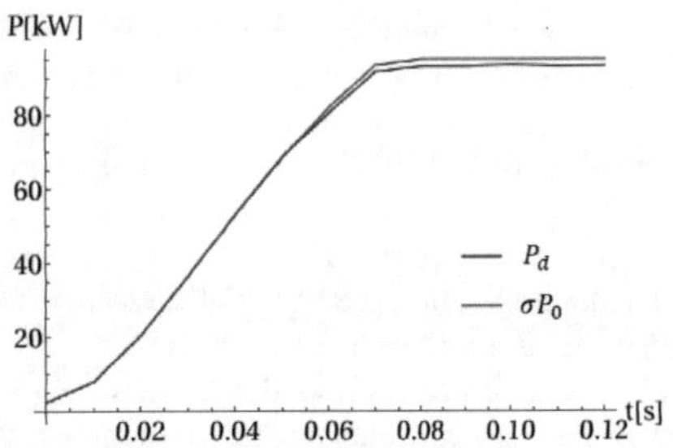

Figure 8. Simulated power absorbed for EBW start-up, calculated from the injected power and optical depth (σP_0) and from the distribution function (P_d).

mains constant once full absorption is reached, as the distribution function remains close to Maxwellian. The effect of collisions might lessen the flattening of the distribution function at high power, and limit the growth of D_0.

The value of D_0 is updated through an iteration, under the criteria that the difference between σP_0 and P_d must be less than 5%. Typically, the value for D_0 that satisfies this criteria is found in no more than three iterations.

The calculated plasma current as a function of time is shown in figure 10, with the fraction carried by energetic electrons (T_e = 50 – 150 keV) also shown. Experiments showed that the plasma current generated by EBW start-up is carried by supra thermal electrons with energies T_e =

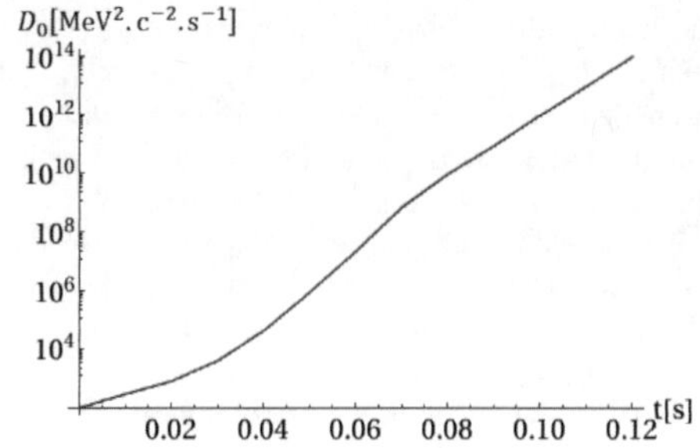

Figure 9. Simulated value for the diffusion coefficient D_0 for EBW start-up.

70 – 100 keV [1, 2], and this is reproduced by simulations, albeit with a wider range of energies.

The preferential heating of electrons creates an asymmetry in both the hot and cold electrons, as shown in figure 11, which carries current in opposite directions. If the loss term were symmetric, this effect would not drive a net current, but due to the asymmetric loss term, electrons with negative $p_{\parallel}$ are lost faster than electrons with positive $p_{\parallel}$. This effect creates a larger population of electrons with positive $p_{\parallel}$, and as most of these electrons are energetic due to the interaction with EBWs, a net current is carried by hot electrons.

Although the generated current is nearly 10 times smaller than the experimentally measured current, resulting in a current drive efficiency of ~ 0.1 A/W rather than 1 A/W, this simple model does produce a population of supra thermal electrons in the same energy range as inferred in the experiments, with these electrons carrying a net current.

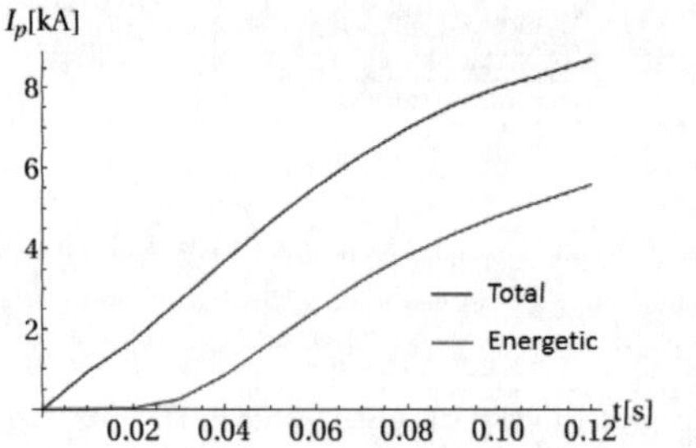

Figure 10. Simulated plasma current for EBW start-up. The total plasma current as well as the current carried only by energetic electrons is shown.

The discrepancy in the energy of electrons carrying the largest fraction of the plasma current can be attributed to the particular choice of parameters. The maximum energy attainable by electrons depend on the EBW heating term, and the specific wave parameters. As illustrated in figure 5, there exists a maximum value of $p_{\perp}$ for which the diffusion term is non-zero, and this value determines the maximum energy attainable by electrons. Beyond this energy, the resonance condition can no longer be satisfied with real values of $p_{\|}$, and the electron can no longer interact with the EBW. A better choice of parameters would limit the energy attainable by these electrons to the same range as observed by experiments.

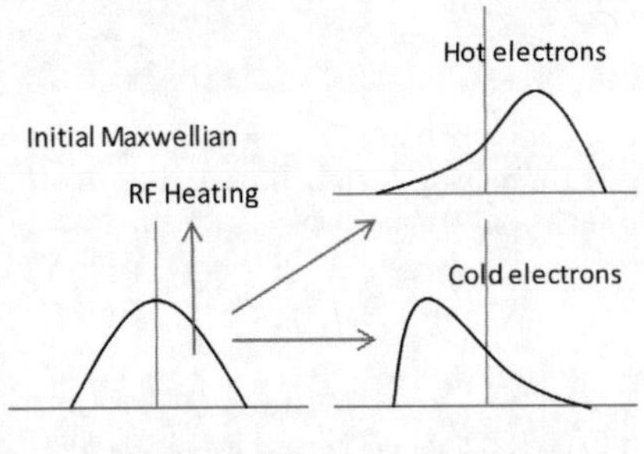

Figure 11. The preferential heating of electrons creates an asymmetry in both the hot and cold electrons, but in different directions.

Inevitable uncertainties and approximations exist in this model, to the point that the choice of a different loss term, for example, may give better quantitative agreement with experiment. The inclusion of additional terms, such as collisions, might also play a role in the calculations, and will have to be included before quantitative comparisons between simulations and experiment can be made.

4 Future Work

Experiments have shown that CFS formed and that a tokamak like equilibrium was established during the EBW start-up phase. Further, a linear dependence between the plasma current and injected RF power was observed. In order to study these effects, the time evolution of the electron density has to be modelled.

Existing, and more complicated, start-up models can be utilised to study the electron density. These models, such as the DYON code [3], rely on the solution of power balance equations to obtain particle densities and temperatures.

For this paper, collisions have been neglected in order to study EBW start-up, as experiments suggest the plasma current is mainly carried by energetic electrons with energies above 70 keV, and these electrons collide very infrequently. This observation was confirmed by simulations, but in order to study EBW start-up under a wider range of conditions, collisions will have to be included and studied.

Lastly, the EBW heating term relies on determining the power absorbed through the optical depth, and equating this to the power absorbed calculated with the distribution function. In this paper, it was assumed that the amount of power converted to EBW was proportional to the electron density, but a more sophisticated method would involve including a ray tracer to calculate the optical depth and conversion to EBW.

5 Conclusion

EBW current drive is a promising method for plasma start-up in spherical tokamaks, for which solenoid-free methods will be needed in burning plasma devices. A kinetic model for studying the time evolution of the electron distribution function is presented, from which the generation of plasma current can be calculated. Experimental evidence suggests that the current generated by EBW current drive is carried predominantly by supra thermal electrons, and this was confirmed by simulations, though the current drive efficiency was less than in experiments. The loss term is asymmetric during start-up, when the magnetic field lines have an open configuration, and this, coupled with a preferential heating of electrons, leads to an asymmetry in the distribution function. The EBW heating accelerates electrons to higher energies by increasing the perpendicular momentum, improving their confinement, to give a current carried by energetic electrons.

A method of relating the spatial dependent plasma wave interaction to a spatially independent diffusion term is also presented. This method relies on equating the volume averaged diffusion term to the absorption efficiency, obtained through knowing the optical depth and amount of power converted to EBW. The diffusion coefficient is chosen such that the power absorbed calculated from the distribution function will equal the power absorbed estimated from the amount of power converted to EBW.

Future studies will include investigating the effect of collisions, the solution of power balance equations to obtain the density evolution, as well as a ray tracer to obtain the wave parameters and absorption efficiency from an arbitrary distribution function.

Acknowledgements

This work was part-funded by the RCUK Energy Programme under grant EP/I501045, the University of York, through the Department of Physics and the WW Smith Fund, and the Golden Key International Honour Society.

References

[1] V.F. Shevchenko *et al.*, *Nucl. Fusion* **50**, 022004 (2010).
[2] V.F. Shevchenko *et al.*, *EPJ Web of Conferences* **87**, 02007 (2015).
[3] H.T. Kim *et al.*, *Nucl. Fusion* **52**, 103016 (2012).
[4] T. Wauters *et al.*, *Plasma Phys. Control. Fusion* **53**, 125003 (2011).
[5] M. Bornatici *et al.*, *Nucl. Fusion* **23**, 1153 (1983).

Study of energetic particle physics with advanced ECEI system on the HL-2A tokamak

Zhongbing Shi[1,*], *Min* Jiang[1], *Liming* Yu[1], *Wei* Chen[1], *Peiwan* Shi[1], *Wulyu* Zhong[1], *Zengchen* Yang[1,2], *Boyu* Zhang[1], *Xiaoquan* Ji[1], *Yonggao* Li[1], *Yan* Zhou[1], *Shaodong* Song[1], *Mei* Huang[1], *Xianming* Song[1], *Jiaxuan* Li[1], *Baoshan* Yuan[1], *Bingzhong* Fu[1], *Zetian* Liu[1], *Xuantong* Ding[1], *Yuhong* Xu[1], *Qingwei* Yang[1], and *Xuru* Duan[1]

[1]Southwestern Institute of Physics, Chengdu, 610041, China
[2]Xihua University, Chengdu, 610039, China

Abstract. Understanding the physics of energetic particles (EP) is crucial for the burning plasmas in next generation fusion devices such as ITER. In this work, three types of internal kink modes (a saturated internal kink mode (SK), a resonant internal kink mode (RK), and a double e-fishbone) excited by energetic particles in the low density discharges during ECRH/ECCD heating have been studied by the newly developed 24(poloidal) × 16(radial) = 384 channel ECEI system on the HL-2A tokamak. The SK and RK rotate in the electron diamagnetic direction poloidally and are destabilized by the energetic trapped electrons. The SK is destabilized in the case of $q_{min} > 1$, while the RK is destabilized in the case of $q_{min} < 1$. The double e-fishbone, which has two m/n = 1/1 modes propagating in the opposite directions poloidally, has been observed during plasma current ramp-up with counter-ECCD. Strong thermal transfer and mode coupling between the two m/n = 1/1 modes have been studied.

1 Introduction

The MHD activities induced by energetic particles (EP), which may cause severe confinement degradation in the burning plasma devices such as ITER and future fusion reactor, have been concerned by many scientists [1]. One of those is the fishbone instability, which was first observed in 1983 in PDX tokamak with nearly perpendicular neutral beam injection (NBI) heating [2]. Then, various types of fishbones, have been observed in many fusion devices [3-8] with different heating, such as electron cyclotron resonance heating and current drive (ECRH/ECCD), lower hybrid current drive (LHCD), and/or ion cyclotron resonance heating (ICRH). The fishbone mode is also called resonant internal kink mode (RK). It is widely accepted that the structure of the RK is a m/n = 1/1 mode, which is destabilized by energetic trapped particles via processional resonance [9]. These phenomena have been explained by resonance between energetic particles and internal kink mode (IKM) at $q = 1$ resonant surface [9,10]. An interesting property of RK is their nonlinear dynamics including robust burst, frequency chirping and jumping, which is associated with the redistribution and the losses of energetic particles. The saturated fishbone mode is also called saturated internal kink mode (SK), which originates from its absence of the $q = 1$ resonant surface (in other words: $q_{min} > 1$) [11]. The SK has characteristics of saturated amplitude, slow change or constant frequency, and long lasting time. Therefore, the SK is also call long lived mode or long lasting mode (LLM) in some literatures. The SKs have been observed and investigated on MAST [12], NSTX [13] and HL-2A [14,15]. When these modes appear, there are strong damping of the toroidal rotation, degradation of stored energy and significant losses of fast particles. Both RK and SK are the core localized internal-kink-like modes, with the mode numbers of m/n = 1/1. Previous studies suggested that, in a weak/reversed shear q profile configuration with a suprathermal electron population, the e-fishbone may be induced in an electron dominated heating plasma.

In this work, three types of internal kink modes excited by energetic particles have been studied by the newly developed electron cyclotron emission imaging (ECEI) system with the ECRH/ECCD heating on the HL-2A tokamak. Significant results are as follows: Both the SK and RK, rotating in the electron diamagnetic direction poloidally, are destabilized by the energetic trapped electrons. The electron SK (e-SK) is destabilized in the case of $q_{min} > 1$, while the electron RK (e-RK) is destabilized in the case of $q_{min} < 1$. A double e-fishbone has been observed for the first time experimentally. It has two m/n = 1/1 modes which propagate in the opposite directions poloidally. The paper is organized as follows: the arrangement of the experiment is presented in section 2. The key diagnostic, 24(poloidal) × 16(radial) = 384 channel ECEI system, is described in section 3. The main experimental results, including the e-RK, e-SK, and double e-fishbone have been shown in section 4 and 5, respectively. The summary is given in section 6.

*Corresponding author: shizb@swip.ac.cn

2 Arrangement of the experiment

The HL-2A device is a medium-sized circular tokamak with a major radius R_{ax} = 165 cm, minor radius a = 40 cm and a closed diverteor, which can be operated in either a limiter configuration or a single null (SN) divertor configuration [16]. The achieved operation parameters are plasma current of I_p = 450 kA, toroidal magnetic field of B_t = 2.7 Tesla, electron and ion temperature of 5 keV and 2.8 keV, respectively, and a line averaged plasma density of ~ 8 × $10^{19}m^{-3}$. The vessel is covered with the carbon fiber compound (CFC) and is routinely siliconized. The available auxiliary heating and current drive power includes 2 MW NBI (ion energy: 45keV), 3 MW ECRH (six gyrotrons, 68 GHz) and 1MW LHCD (two klystrons: 2.45 GHz). The ECRH power with O mode (selected fundamental harmonic) or X-mode (second harmonic) is injected from the low field side (LFS). The transmission efficiency can achieve 0.9. The launching angle of ECRH antennas can vary from -30 to 30 degree, toroidally. This enable us to change the driving effciency and the driving direction.

To study the instability of energetic particles, the experiments are arranged at low density condition (about 0.5 × $10^{19}m^{-3}$) with high auxiliary heating power, in order to achieve high temperature but low collisionality operation regime. The previous work suggested that the internal kink mode, i.e. fishbone, can be easily excited in a weak/negative magnetic shear configuration [14, 15]. The electron cyclotron current drive is used to control the magnetic configuration. Four gyrotrons are used in this experiment. Among them, two gyrotrons are used for heating and the other two are used for electron cyclotron current drive (ECCD).

The key diagnostics for this experiment are the soft X-ray array, the far-infrared (FIR) laser interferometer, the X-mode reflectometer, the ECE and ECEI diagnostic. The soft X-ray intensity is sensitive to the electron temperature, density and effective charge, and is given by $I_{sxr} \sim n_e^2 T_e^a f(Z_{eff})$ [17]. The exponent a is an increasing function of the absorber foil thickness; its value is usually between 2 and 3. Therefore, small perturbations of n_e, T_e and Z_{eff} can easily be detected. The multichannel far-infrared interferometer-polarimeter is used to measure the line averaged electron density and the Faraday rotation angle [18]. The Q/V band X-mode reflectometer is used to measure the density profile with the temporal resolution of ~ 0.02ms, in the case of B_t ~ 1.3 Tesla [19]. The 56-channel ECE with spatial resolution about 1 cm is used to measure the temperature profile and its evolution across the whole plasma area [20]. The absolute value of the electron temperature is obtained by cross calibration with temperature measured by the core Thomson Scattering system. It should be noted that the FIR laser interferometer and soft-x-ray record central chord line-averaged signals and chord-integrated signals, respectively, whereas the X-mode reflectometer and ECE/ECEI diagnostic record local parameters of electron density and temperature in concert with the reduced toroidal magnetic field, which decreases monotonically with radius. Due to its local measurement and high spatial resolution, the ECE/ECEI diagnostics are also used to determine the accurate locations and the mode structures of the MHD instabilities. The local temperature measurement from ECE/ECEI diagnostic is determined by the optical thickness, which has been discussed in many literatures [20, 21]. In this experiment, the electron densities of the analyzed discharges are about 0.5 × $10^{19}m^{-3}$. This will cause optical thin at the plasma edge. But at the plasma center, inside r/a = 0.5, the calculated optical thickness is tau > 2 in our analyzed discharges. Since the observed internal kink modes are located inside r/a = 0.5 in this work, the ECE/ECEI measurements are reliable.

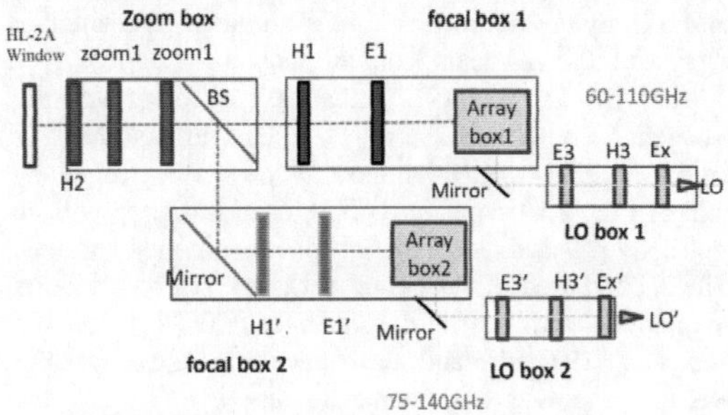

Fig.1. Schematic diagram of the dual-array ECEI optical system on HL-2A.

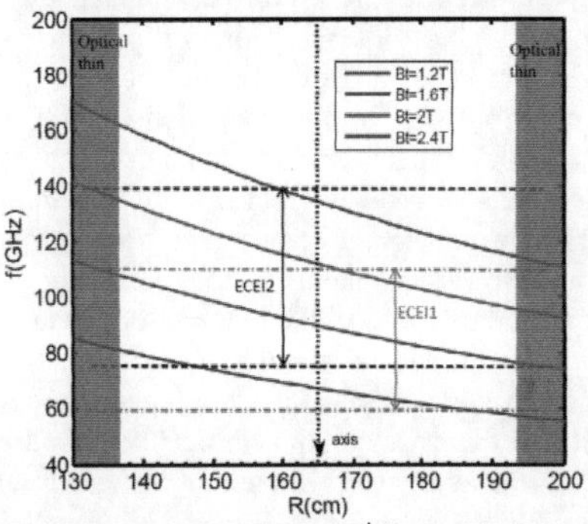

Fig.2. Frequency range of the 2nd harmonic X-mode ECE radiation as a function of HL-2A major radius. The shaded area denotes the third harmonic overlap region, where the optical thickness is thin.

3 384 channel ECEI system

The electron cyclotron emission imaging (ECEI) system on the HL-2A tokamak has been developed to visualize the two-dimensional (2D) temperature fluctuations. The ECEI system is comprised of a quasi-optical system, a front-end 24 channel heterodyne imaging array with a tunable radio frequency from 60 to 140 GHz, and a set of back-end ECEI electronics that together generate 24(poloidal) ×16(radial) = 384 channel images of the 2D temperature fluctuations.

Figure 1 shows the schematic diagram of the ECEI optical system. It is a dual array ECEI system. The optical system consists of one zoom optics box, two focal optics boxes and two local oscillator (LO) optics boxes. The zoom optics box has two E-plane zoom lenses, and the field of view can be adjusted easily by changing the position of the zoom lenses. The ECE signal from plasma is captured by the zoom lenses. Then, the signal is power divided by a 3 dB beam splitter (BS). One part of the beam transmits through the BS to the focal optics box 1. The beam goes through the focal lenses H1 and E1, arrives at the first antenna array in the focal optics box 1, and is then down-converted to the intermediate frequency (IF) by the microwave mixers in the first antenna array with the LO beam from the first LO optics. Another part of the beam is reflected by the BS, then reflected by a large mirror, goes through the focal lenses H1 and E1 in the focal optics box 2, arrives at the second antenna array, and is then down-converted to the IF by the microwave mixers in the second antenna array with the LO beam from the second LO optics.

The two LO and focal optics have different working frequencies. One works at 60-110 GHz and the other one works at 75-140 GHz. Figure 2 shows the frequency range of the 2nd harmonic X-mode ECE radiation with B_t = 1.2 - 2.4 Tesla as a function of HL-2A major radius. The shaded area in the figure marks the region which is overlapped with the third harmonic ECE (R < 137 cm and R > 193 cm) and the optical thickness is thin. Because the two LO frequencies are independent, the measurement region can be flexibly adjusted according to the physical requirement, by changing the working frequency under different toroidal magnetic field operations. The size of the working pattern, or the vertical resolution, is determined by the zoom optics. By changing the position of the zoom lenses, the optical system can work at between wide and narrow zoom. The wide zoom has a vertical resolution of 2.3 cm. It is designed to provide a large viewing coverage of the plasma with a vertical height of 55 cm. This is very useful for the large-scale MHD structure measurement. The narrow zoom has a vertical resolution of 1.7 cm. It is designed to view the delicate structures of the small-scale MHD in the plasma with a vertical height of 40 cm. Details of the system calibration can be found in reference [22].

4 Resonant internal kink mode and saturated internal kink mode

The internal kink mode induced by energetic electrons can be excited easily during high power on-axis ECRH in the low density plasmas on HL-2A [7,14,23]. The magnetic shear can be controlled by the current drive. The current driving direction of ECCD can change the magnetic shear, which affect the excitation of the internal kink mode. In order to clarify the effect of heating and driving direction of EC wave on internal kink mode, two discharges with opposite ECCD driving directions have been studied. As shown in Fig.3(a) and (c), they have same plasma current, density and ECRH/ECCD power, which are 133 kA, 0.5 × 10^{19} m $^{-3}$ and 1 MW, respectively. Figure 3 (b) and (d) show spectrograms of the central soft X-ray signals. The e-RK is found in condition of counter-ECCD/ECRH. The frequency of the e-RK decreases from 15 to 5 kHz and accompanied with sawtooth crashes. The e-SK is found in the condition of co-ECCD/ECRH. Unlike conventional RK with frequency chirping or jumping behaviors [7,14], its frequency is nearly unchanged. This suggests that this mode is saturated. It may closely relate to the weakly reversed shear due to ECCD [24,25].

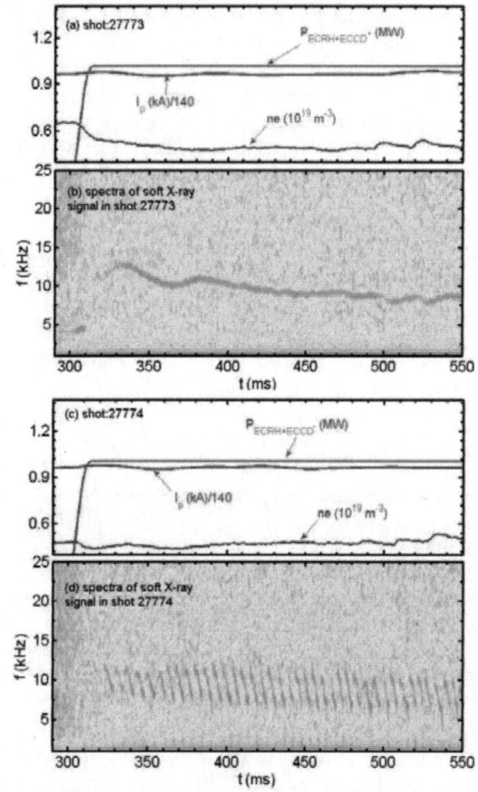

Fig. 3. Effects of current driving directions of ECCD on the internal kink mode. (a,c) discharge parameters, and (b,d) spectrograms of soft X-ray signals in shot 27773 and 27774.

Figure 4 shows time evolution of the mode structures of the e-RK and e-SK measured by the ECEI system. There are two obvious islands, a hot and a cold one, and the mode locates almost within a r = 10 cm circle (the q = 1 surface). So the mode numbers are m/n = 1/1. The intensity of e-SK is higher than that of e-RK. This suggested that the SK might cause a strong damping of the toroidal rotation, degradation of stored energy and significant fast particle loss. Though the opposite current driving directions of ECCD for e-RK and e-SK, their propagating directions are in the anticlockwise direction, which is the electron diamagnetic direction, poloidaly.

Figure 5 shows the q-profiles during e-RK and e-SK modes in shot 27773 and 27774. They are calculated by the current filament method and modified by Faraday rotation measured by FIR laser polarimeter interferometer. The negative q-profiles are formed by the high power ECRH for both conditions. The q_{min} is a slight bigger than unity with co-ECCD at t = 505 ms in shot 27773, while the q_{min} is a bit smaller than unity with counter-ECCD at t = 505 ms in shot 27774. There are obvious sawtooth crashes in shot 27774 (RK condition) from soft X-ray sig nals, while there are no sawtooth in shot 27773 (SK condition). The signals from soft X-ray arrays also indicate that the q_{min} is a little bigger than unity in shot 27773, while in shot 27774, the q_{min} is smaller than unity. This suggests that there is a q_{min} condition for the destabilization of SK. The SK is destabilized in the case of $q_{min} > 1$, while the RK is destabilized in the case of $q_{min} < 1$.

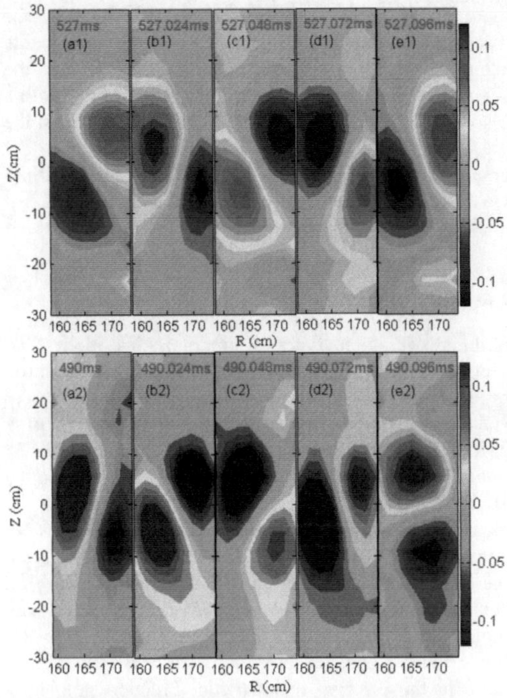

Fig.4. Mode structure evolutions of e-RK (a1,b1,c1,d1,e1) in shot: 27774 and e-SK (a2,b2,c2,d2,e2) in shot: 27773 measured by the ECEI system.

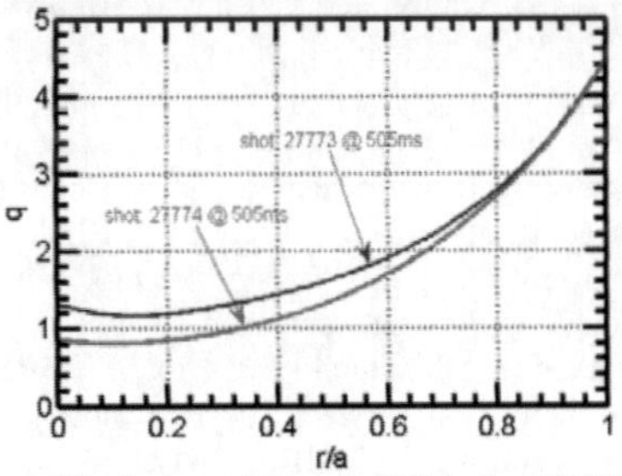

Fig. 5. q-profiles during e-RK and e-SK in shot 27773 and 27774 at t = 505 ms.

5 Double e-fishbone

With counter-ECCD injection during the plasma current ramp-up, the double structure of e-fishbones have been observed [26]. The typical discharge with double e-fishbone instability is shown in Fig.6. Figure 6(a) shows the plasma current, which ramps up from 146 kA at 540 ms to 165 kA at 590 ms. The counter-ECCD/ECRH with power of 850 kW was injected into plasma during current ramp up. As shown in Fig.6(c) and (d), evident electron temperature fluctuation appears at 548 ms, and its amplitude gradually increases and frequency gradually drops from 20 kHz to 10 kHz until 563 ms. The mode disappears at 563 ms due to the gas puffing, which causes the increase of the density at this time. The decrease of the mode frequency may be due to the variation of the q_{min}. After about 5 ms, when the density back to about $0.5 \times 10^{19} m^{-3}$, the mode appears again, and the mode frequency nearly maintains at about 7.5 kHz from 568 ms to 579 ms .

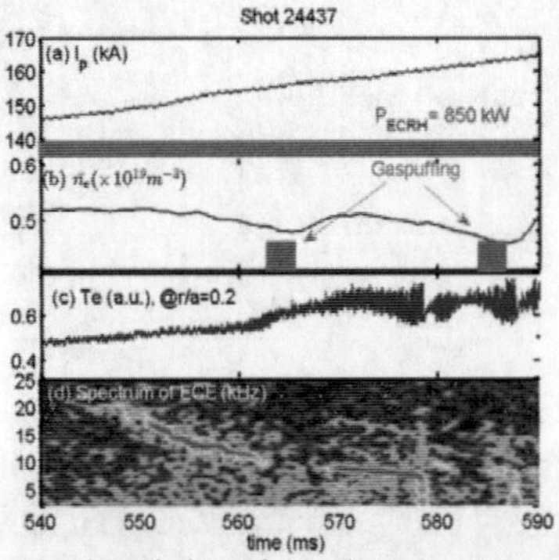

Fig.6. Temporal evolution of (a) plasma current and ECRH power (b) line-averaged electron density and gas puffing pulses, (c)ECE signal at r/a=0.2, (d) Frequency spectrum of central ECE signal.

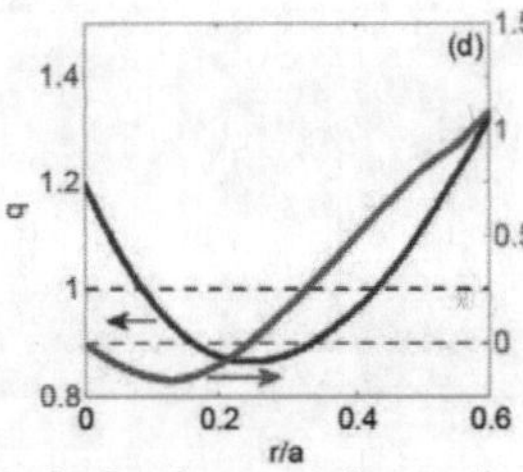

Fig.7. Profiles of safety factor q and magnetic shear s in shot 24437 at 570 ms.

The deposition of the ECRH and amplitude of driven current are simulated through a linear ray tracing code TORAY-GA [27], which computes the trajectories of Gaussian rays representing ECRH microwave beams using cold plasma dispersion relation. The ECRH power is almost deposited inside $r/a = 0.35$, and peaks at $r/a = 0.2$. The driven current is about 13.6 kA. Figure 7 shows the profiles of the safety factor q and the magnetic shear s at 570 ms in shot 24437. The q-profile is constructed by the current filament method, where the plasma equilibrium configurations are fitted by using the internal magnetic field measurements of Faraday rotation effect, and external magnetic field measurements with magnetic probes and flux loops. The magnetic shear s is given by $s = rdq/qdr$. We can find that there are two $q = 1$ surfaces. The locations of the two $q = 1$ surfaces are $r/a = 0.12$ and $r/a = 0.42$ respectively. It is observed that

DOI: 10.1051/epjconf/201714701003

there is weak negative shear configuration in the core plasma.

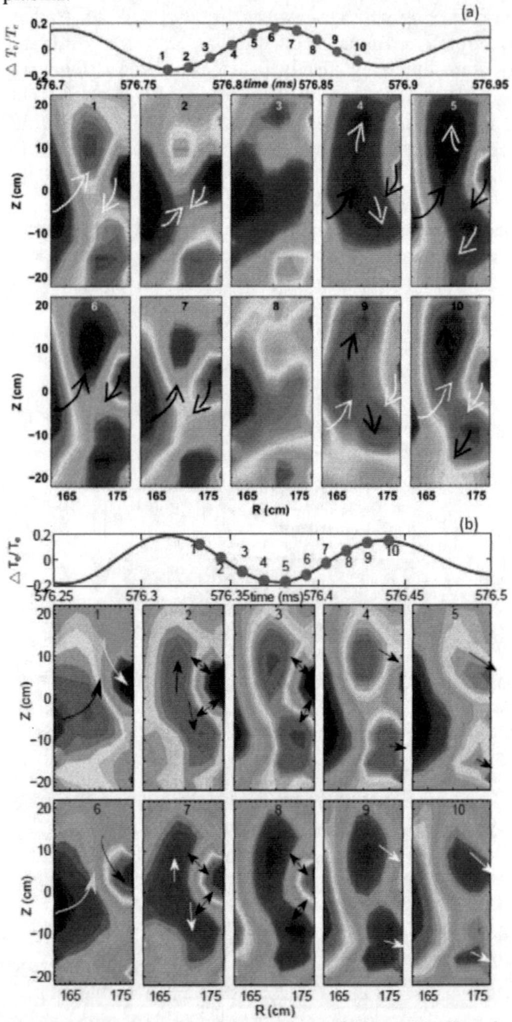

Fig.8. Temporal frames of the double e-fishbone (a) when the hot-hot (cold-cold) spots of the two modes collide with each other, (b) when the hot (cold) spot of one mode approaches the cold (hot) spot of the other mode. The white and black arrows indicate the moving trend of the cold and hot spots, respectively. The double arrow means the interaction between the two modes. When hot spots meet with cold spots, the enhanced outwards transport was observed.

The temporal evolutions of the 2D structures of the double e-fishbone have been observed by ECEI system. It has two m/n = 1/1 modes. The locations of the two modes are found to be R = 170 cm (r/a = 0.15) and R = 178 cm (r/a = 0.41), which are consistent with that of the two q = 1 surfaces shown in Fig.7. As shown in Fig.8, the two m/n = 1/1 modes with opposite propagation directions have been found clearly: the inner mode propagates in the electron diamagnetic drift direction and the outer mode propagates in the ion diamagnetic direction. In these frames, we can see that there are two cold spots and two hot spots with different propagation direction respectively. The white and black arrows indicate the moving trend of the cold and hot spots, respectively. As seen in Fig. 8(a), when the two cold (or hot) spots are far away, they can be clearly recognized to rotate in the opposite direction. When they are close to each other, the two clusters of cold (or hot) spots seems to merge together and then becomes elongated. Afterwards, the elongated cold spots seems to split into two parts, and continue to move in the opposite directions. In this dynamics process, no enhanced transport has been observed. However, as seen in Fig. 8(b), when the inner hot spot meets the cold spot of the outer mode, the obvious heat energy transition has be observed between the hot and cold spot. The double arrow in this figure denotes the interaction between the two modes. This process can be observed clearly with a cartoon or movie. The heat transfer seems to begin at the largest gradient of the temperature, where the temperature decreases firstly, then the hot spot splits into two parts and the two heat parts transfer outward quickly.

6 Summary

In this work, three types of internal kink modes (an electron saturated internal kink mode, an electron resonant internal kink mode (RK), and a double e-fishbone) excited by energetic particles during ECRH/ECCD heating have been studied by the 24 (poloidal)×16 (radial) = 384 channel ECEI system on the HL-2A tokamak. These modes have been found in the low density discharges and their frequencies are close to the precessional frequency of the trapped fast electrons. Their mode numbers and mode structures have been identified by the newly developed ECEI system. The frequency of e-SK is nearly unchanged, while the e-RK shows rapid frequency chirping. The e-SK and e-RK rotate in the electron diamagnetic direction poloidally and are destabilized by the energetic trapped electrons. We found that there is a q_{min} threshold for the destabilization of e-SK. The e-SK is destabilized in the case of $q_{min} > 1$, while the RK is destabilized in the case of $q_{min} < 1$. The double e-fishbone has two m/n = 1/1 modes which propagate in the opposite directions poloidally. The two modes are excited in a weak/negative magnetic central shear configuration, which is achieved with localized ECRH/ECCD during plasma current ramp-up. Strong thermal transfer and mode-mode coupling between the two m/n = 1/1 modes have been observed, which seems to be related to the nonlinear interaction between the two modes and local heat transport during the MHD activity.

This work is supported by Chinese National Fusion Project for ITER under Grant No.2013GB104000, 2014GB107001 and 2013GB107002, and National Natural Science Foundation of China (11275062, 11475057, 11405049, 11475058, 11005037).

References

1. A. Fasoli, et.al., Nucl. Fusion **47**, S264 (2007)
2. K. McGuire, et.al., Phys. Rev. Lett.**50**, 891 (1983)
3. W. Chen, et.al., Nucl. Fusion **50**, 084008 (2010)

4. X. T. Ding, et.al., Nucl. Fusion **42**, 491 (2002)
5. L. W. Yan, et.al., Chin. Phys. Lett. **18**, 1227 (2001)
6. Macor, et.al., Phys. Rev. Lett. **102**, 155005 (2009)
7. L. M. Yu, et. al., Nucl. Fusion **53**, 053002(2013)
8. F.Nanais, et.al., Nucl. Fusion **45**, 102509 (2005)
9. L. Chen, et. al., Phys. Rev. Lett. **52**, 1122 (1984)
10. R. Betti, et.al., Phys. Rev. Lett. **70**, 3428 (1993)
11. F. Wang, et.al., Phys. of Plasmas **20**, 102506 (2012)
12. I.T. Chapman, et.al., Nucl. Fusion **51**, 073040 (2011)
13. J.E. Menard, et.al., Nucl. Fusion **45**, 539 (2005)
14. W. Chen, et.al., Nucl. Fusion **49**, 075022 (2009)
15. W. Deng, et.al., Nucl. Fusion **54**, 013010 (2014)
16. X.R. Duan, et. al., Nucl. Fusion **53**, 104009 (2013)
17. Equipe TFR, Nucl. Fusion **18**, 647 (1978)
18. Y. Zhou, et.al., Rev. Sci. Instrum. **83**, 10E336 (2012)
19. W.L. Zhong, et.al., Rev. Sci. Instrum. **85**, 013507 (2014)
20. Z. B. Shi, et.al., Rev. Sci. Instrum. **85**, 023510 (2014)
21. M. Bornatici, et.al., 1978 Nucl. Fusion **23**, 1153 (1983)
22. M. Jiang, et.al., Rev. Sci. Instrum. **84**, 113501 (2013)
23. L. M.Yu, et.al., Nucl. Fusion, **57**, 036023 (2017)
24. T. Chapman, et.al., Nucl. Fusion **50**, 045007 (2010)
25. R. B. Zhang, et. al., Plasma Phys. Control. Fusion **56**, 095007 (2014)
26. M. Jiang, et.al., Phys. Plasmas, **24**, 022110 (2017)
27. K.Matsuda, et.al., IEEE Trans. Plasma Sci. **17**, 6 (1989)

Summary of ECE Presentations at EC-19

M. E. Austin[1,*]

[1]Institute for Fusion Studies, University of Texas, Austin, TX 78712, USA

Abstract. At the EC-19 workshop there were ten presentations in the topic of diagnosing and heating plasmas with Electron Cyclotron Emission (ECE) and Electron Bernstein Wave (EBW). Innovations continue in designs of ECE systems while tried-and-true instruments still provide adequate data to explore new areas of research. Heterodyne radiometers and Michelson interferometers carry on as the bedrock of ECE electron temperature (T_e) measurements while ECE imaging systems continue expanding their capabilities. The design of the ITER-ECE diagnostic system is in its advanced stages; solutions are being found for challenges in the transmission lines and front-end calibration sources.

1 Introduction

The ECE topics at EC-19 were presented in three talks and seven posters. The topics naturally divide themselves into the primary type of instrument involved, Michelson interferometer, heterodyne radiometer, and ECE-Imaging instruments, with a separate section for ITER ECE systems plus a special related topic of EBW heating.

2 Measurement systems

2.1 Measurements with Michelson interferometers

Two papers dealt with the calibration of Michelson interferometers. A. Sinha reported on the calibration of the Michelson that will be used on the SST-1 tokamak. The system was calibrated at Culham before being sent to India. Two blackbody sources at 873 K and 77 K respectively were used, obtaining well-resolved response curves and verifying the stability of the instrument. M. Austin showed measurements of the variation of sensitivity of the Michelson instrument with humidity. For the DIII-D tokamak's system with the 18 m corrugated waveguide transmission line in air the results showed ignorable changes in response for low harmonic frequencies (70-300 GHz), less than 1% for relative humidity changes of 25% to 67%. But attenuation due to water vapor increases strongly with frequency such that the changes brought about at high frequency (300-1000 GHz) can be 5% or greater and hence need to be taken into account for data used in this range.

* Corresponding author: max.austin@utexas.edu

2.2 Measurements with heterodyne radiometers

Three papers showed results from heterodyne systems. V. Siju summarized the radiometer diagnostics on the IPR tokamaks. The SST-1 system, eight channels covering 75-85 GHz, is providing well resolved T_e data. The ADITYA-U system is under construction; it will consist of 16 channels covering 75-94 GHz. S. Denk gave a talk discussing measurements in ASDEX-U of non-thermal ECE in the edge channels of that tokamak's 60 channel system. Forward modeling of the edge emission data has provided valuable information about the non-thermal distribution. Finally, M. Fontana presented data from the TCV tokamak heterodyne radiometer suite. TCV has multiple systems with a variety of flexible configurations. A highlight of the data was from TCV's Correlation-ECE system, that showed a strong reduction in T_e fluctuations for plasma discharges with a negative triangularity shape over an equivalent positive triangularity shape, indicating a causal relationship between the reduced turbulence and the improved confinement in the "backwards-D" configuration.

2.3 Measurements with ECE-Imaging

Results were presented from the ECE-Imaging system on KSTAR by M. Choi. The 2-D system was used to observe the onset of m/n=2/1 magnetic islands with various actuators. In the case of a density ramp actuation the 2-D contours closely resemble ballooning mode simulations. Mode structure was mapped out during the growth, crash, and recovery phases of both small and large minor disruptions. KSTAR now has a second ECE-I system located 22.5 degrees toroidally from the first system; this permits 3-D imaging of plasma T_e structures. This capability made possible observation of the modes of major disruption where the 2/1 magnetic island was seen to interact with a cold bubble which affected its growth. This could be one explanation for the explosive 2/1 island growth in a major disruption.

2.4 ITER ECE systems

There were three papers discussing ECE systems related to ITER. S. Danani gave a talk on the testing of a prototype receiver for the ITER high frequency radiometer. The receiver, which covers the 200-300 GHz range, was put through noise temperature and sensitivity checks and subsequently measured plasma signals at the DIII-D facility. The instrument met ITER requirements of 20,000 K noise temperature and 500 V/W sensitivity and was shown to make linear, frequency resolved measurements of plasma T_e when compared to simultaneous data from the DIII-D Michelson interferometer. R. Kumar presented data from the characterization of a high temperature blackbody source intended to be used in the ITER instrument room. IR measurements at 500 ˚C validated the high emissivity and good thermal conductivity of the silicon carbide source element.

G. Taylor gave an overview talk on the status ITER ECE diagnostic system. Good progress has been made in the design of interlocking systems that are shared between the US and IN-DA groups. The front end optics design is nearly final; recent changes were made to redirect the oblique view away from the ECH launchers and to accommodate improved neutron shielding. The hot calibration source (1000 K operating temperature) has been through several design iterations in order to meet tough requirements for shock loads, radiation loads, and overall robustness. The path of the transmission line (~45 m length) and the beam modification units (splitter boxes) are designed. The final details of the transmission line waveguide are awaiting testing of the two candidate options, smooth wall or dielectric wall, both which would be in circular metal tubing. The layout of the ITER ECE instrument hall, containing two Michelson interferometers (70-1000 GHz), a low frequency radiometer (122-230 GHz) and a high frequency radiometer (220-340 GHz) has been decided.

2.5 EBW heating

A. Koehn presented a paper on electron Bernstein wave (EBW) heating on the TJ-K stellarator. EBW techniques are a way to launch and receive waves in over-dense plasmas and in this case the heating and current drive aspects were investigated. Using a novel phased array antenna that steers via small frequency changes, heating was observed as the density was increased to the point where the O-X-B mode conversion for 2nd harmonic took place. Additionally, reduction of broadband turbulence was obtained when heating at high harmonic frequencies.

3 Conclusions

ECE continues to be a major diagnostic for T_e measurements in magnetically confined high temperature plasmas. Heterodyne radiometers and Michelson interferometers are still the major contributors for most experiments. Advanced versions of heterodyne instruments, such as correlation radiometers and 2- and 3-D imaging systems are exploring the finest details of electron temperature structures in plasmas. The design of the ITER-ECE diagnostic system is well along. Challenges are being found due to the machine size and nuclear aspects of ITER, particularly in the areas of hot source development, transmission line design. These challenges are being met and it is expected that the ECE diagnostic will be an important contributor to ITER plasma physics understanding.

EPJ Web of Conferences **147**, 02002 (2017) DOI: 10.1051/epjconf/201714702002
EC-19

Radiation transport modelling for the interpretation of oblique ECE measurements

Severin S. **Denk**[1,2], Rainer **Fischer**[1], Omar **Maj**[1], Emanuele **Poli**[1], Jörg K. **Stober**[1], Ulrich **Stroth**[1,2], Branka **Vanovac**[3], Wolfgang **Suttrop**[1], Matthias **Willensdorfer**[1], and the ASDEX Upgrade Team[1]

[1] *Max Planck Institute for Plasma Physics, Boltzmannstr. 2, 85748 Garching, Germany*
[2] *Physik-Department E28, Technische Universität München, 85748 Garching, Germany*
[3] *FOM-Institute DIFFER, Dutch Institute for Fundamental Energy Research, 5600 HH, Eindhoven, The Netherlands*

Abstract. The electron cyclotron emission (ECE) diagnostic provides routinely electron temperature (T_e) measurements. At ASDEX Upgrade an electron cyclotron forward model, solving the radiation transport equation for given T_e and electron density profile, is used in the framework of integrated data analysis. With this method T_e profiles can be obtained from ECE measurements even for plasmas with low optical depth. However, due to the assumption of straight lines of sight and an absorption coefficient in the quasi-perpendicular approximation this forward model is not suitable for the interpretation of measurements by ECE diagnostics with an oblique line of sight.
Since radiation transport modelling is required for the interpretation of oblique ECE diagnostics we present in this paper an extended forward model that supports oblique lines of sight. To account for the refraction of the line of sight, ray tracing in the cold plasma approximation was added to the model. Furthermore, an absorption coefficient valid for arbitrary propagation was implemented. Using the revised model it is shown that for the oblique ECE Imaging diagnostic at ASDEX Upgrade there can be a significant difference between the cold resonance position and the point from which most of the observed radiation originates.

1 Introduction

For the measurement of both, the electron temperature (T_e) profile and T_e fluctuations, Electron Cyclotron Emission (ECE) is a widely used diagnostic. However, every ECE diagnostic inherently measures radiation temperature (T_{rad}) and not T_e. For many cases it is possible to assume that fluctuations in T_{rad} are identical to T_e fluctuations at cold resonance. But this procedure fails if absorption is low near the cold resonance and/or if the line of sight (LOS) is not perpendicular to the magnetic field lines. If the LOS of ECE diagnostic is not perfectly perpendicular to the magnetic field, the Doppler effect leads to a broadening of the local cyclotron frequency. If the ECE diagnostic is located on the low field side (LFS), the down-shifted emission can be reabsorbed close to the cold resonance, but the up-shifted emission cannot. Hence, the location of the measurement is not necessarily the cold resonance position. In this article an electron cyclotron forward model is introduced that allows the determination of the region in the plasma that contributes to the measured T_{rad}.

The article is structured as followed. In the second section that follows this introduction the forward model is introduced. Afterwards, in the third section, the limitations of the model are discussed. Finally, in the forth section. an example for the application of the forward model is given. Conclusions are drawn at the end.

2 Electron cyclotron forward model

The purpose of the electron cyclotron forward model is to calculate T_{rad} for a given antenna pattern of the diagnostic, the T_e profile, the electron density (n_e) profile and the magnetic equilibrium. The process of calculating T_{rad} can be decomposed into two steps. First ray tracing is performed to obtain the LOS of the diagnostic and, second, the radiation transport along the LOS to obtain T_{rad} is calculated.

2.1 Ray tracing in the limit of geometrical optics

The implementation of ray tracing in the present model closely follows ref. [1]. The only distinction is that diffraction is neglected. Differently from ref [1], a single ray, rather then a whole diffracting beam, is traced, hence only standard ray tracing equations are solved [2]:

$$\frac{\mathrm{d}\vec{x}}{\mathrm{d}s} = \left.\frac{\partial\Lambda/\partial\vec{N}}{\left|\partial\Lambda/\partial\vec{N}\right|}\right|_{\Lambda=0} \qquad \frac{\mathrm{d}\vec{N}}{\mathrm{d}s} = -\left.\frac{\partial\Lambda/\partial\vec{x}}{\left|\partial\Lambda/\partial\vec{N}\right|}\right|_{\Lambda=0} \tag{1}$$

The arclength is given by s and $\vec{x}$ is the spatial coordinate. The second canocical coordinate $\vec{N} := \frac{c_0\vec{k}}{\omega}$ with $\vec{k}$ the wave vector, ω the angular (measured) frequency and c_0 the vacuum speed of light. With Λ the cold plasma dispersion relation:

$$\Lambda(\vec{x}, \vec{N}, \omega) = |\vec{N}|^2 - N^2_{s,\omega}(X, Y, N_\parallel) \tag{2}$$

and $N_{s,\omega}$ is the cold plasma refractive index:

$$N_{s,\omega}(X, Y, N_\parallel) = 1 - X - XY^2 \frac{1 + N_\parallel^2 \pm \Delta}{2(1 - X - Y^2)} \quad (3)$$

Where "+" corresponds to the extraordinary mode (X-mode) and "−" corresponds to the ordinary mode (O-mode).

$$\Delta := \sqrt{(1 - N_\parallel^2)^2 + 4N_\parallel^2 \frac{1 - X}{Y^2}} \quad (4)$$

In equation (2) $X := \frac{\omega_p^2}{\omega^2}$ has been introduced with $\omega_p = \sqrt{\frac{n_e e^2}{\epsilon_0 m_{e,0}}}$ the plasma frequency and e is the elementary charge. The vacuum permittivity is symbolized as ϵ_0 and $m_{e,0}$ the electron rest mass. Furthermore, the normalized cyclotron frequency $Y := \frac{\omega_{c,0}}{\omega}$ was introduced in equation (2), where $\omega_{c,0} = \frac{eB}{m_{e,0}}$ is the cyclotron frequency and B the absolute magnetic field strength. The component of $\vec{N}$ that is aligned with the magnetic field line is denoted as $N_\parallel := |\vec{N}| \cos\theta := \frac{\vec{N}\cdot\vec{B}}{|\vec{B}|}$. The component perpendicular to the magnetic field lines is analogously defined to be $N_\perp := |\vec{N}| \sin\theta$. For $\vec{B}$ the magnetic field ripple of ASDEX Upgrade is considered [3].

The two equations in (1) form a system of two coupled three dimensional differential equations. They are solved numerically using the DLSODE [4] algorithm. The derivatives in equation (1) are expanded using chain rules, which allows analytical evaluation of most the derivatives. Only the spatial derivatives of the numerical input quantities (n_e profile, magnetic field and equilibrium) are performed numerically. For all interpolations the spline routines given by ref. [5] are used. The splines also provide all the required derivatives. The problem of the transition from vacuum propagation to plasma propagation is performed using root solving accounting also for the additional refraction on plasma entry via Snell's law. The first point of plasma propagation is given by the outermost point given by the input density profile.

To assure that the implementation of ray tracing is correct, the present model was benchmarked against the TORBEAM code [6]. Although TORBEAM does account for diffraction, the central ray of a TORBEAM calculation is only subject to refraction. In the following, one of the benchmarking scenarios is discussed.

For this particular case the LOS is reflected near the plasma core. As an example a channel of the slightly oblique ECE Imaging system (ECEI) array [7] at ASDEX Upgrade was chosen (s. a. section 4). As plasma scenario #30839 at t = 1.75 s was selected. In the top graph of figure 1 the poloidal cross section of ASDEX Upgrade is shown. Flux surfaces are depicted as black dashed lines and the separatrix is highlighted with a blue solid line. In the bottom of figure 1 the top view of ASDEX Upgrade is shown. In both graphs the LOS according to the presented model (green) and TORBEAM [6] (dashed purple) are shown. On the entire LOS up shortly before the reflection the two rays do not deviate by more than 2 mm. This deviation is expected when comparing ray tracing codes,

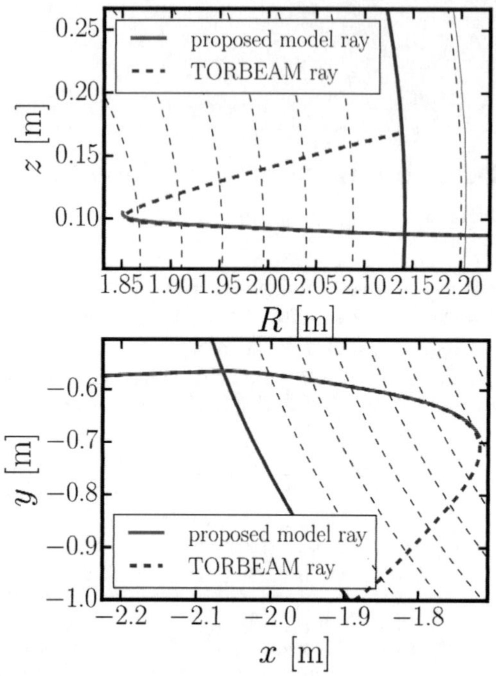

Figure 1. The top graph shows the poloidal cross section of ASDEX Upgrade for # 30839 at t = 1.75 s. The bottom graph shows the corresponding top view. Both graphs show one of the LOSs of the ASDEX Upgrade ECEI system. The green LOS is obtained by the present model and the purple LOS was calculated by the TORBEAM code (s. ref. [6]).

especially if refraction is strong (s. a. ref. [8]). But even this small deviation has a strong effect on the reflection, showing how strongly the reflection is altered by slight changes in the ray properties. For this reason propagation is stopped in the present model upon a sign change in $\frac{\partial|\vec{B}|}{\partial s}$. Therefore, multiple passes through a resonant layer by an internal reflection are explicitly excluded from the model. A direct consequence of this is that an internal reflection by the plasma is treated like a wall reflection with the corresponding inaccuracies. A description of the model for wall reflections can be found in section 2.3. In section 3.1 it is discussed, how reflections (both by the wall or by the plasma) affect the accuracy of the model.

2.2 Radiation transport and EC absorption

In this section the modelling of the radiation transport is explained. Only a single ray and measured frequency are considered. Hence, for each measured frequency the radiation transport equation has to be solved only once. The radiation transport equation (5) is given as [9]:

$$\frac{\mathrm{d}}{\mathrm{d}s} \frac{I_\omega(s)}{N_{\omega,\mathrm{ray}}^2(s)} = \frac{1}{N_{\omega,\mathrm{ray}}^2(s)} \left(j_\omega(s) - \alpha_\omega(s) I_\omega(s)\right) \quad (5)$$

Integrating equation (5) over the LOS yields the intensity $I_\omega(s_{\mathrm{ant}})$ at the antenna position s_{ant}. Furthermore, s is the LOS coordinate, $N_{\omega,\mathrm{ray}}$ the ray refractive index , $j_\omega(s)$ the emissivity and $\alpha_\omega(s)$, where introduced in equation (5). Although the "ray refractive index" $N_{\omega,\mathrm{ray}}$ appears in the radiation transport equation it can be eliminated with a simple substitution $\frac{I_\omega(s)}{N^2_{\omega,\mathrm{ray}}(s)} \to \Xi_\omega(s)$:

$$\frac{\mathrm{d}}{\mathrm{d}s}\Xi_\omega(s) = \frac{j_\omega(s)}{N^2_{\omega,\mathrm{ray}}(s)} - \alpha_\omega(s)\Xi_\omega(s) \tag{6}$$

Solving equation (6) yields $\Xi_\omega(s_{\mathrm{ant}})$. At the antenna position, however, there is no plasma and $N^2_{\omega,\mathrm{ray}}(s_{\mathrm{ant}}) = 1$ and $\Xi_\omega(s_{\mathrm{ant}}) = I_\omega(s_{\mathrm{ant}})$. Hence, $N^2_{\omega,\mathrm{ray}}(s)$ is not needed in the radiation transport equation as long as $\frac{j_\omega(s)}{N^2_{\omega,\mathrm{ray}}(s)}$ is used in place of $j_\omega(s)$. The radiation transport equation can be solved by a conventional, explicit Runge-Kutta scheme (e.g. Rk4). Both, the absorption coefficient and the emissivity are strongly peaked near the cold resonance position. Hence, it is useful to choose a smaller step size close to the cold resonance [10].

To solve the radiation transport equation both, the emissivity $j_\omega(s)$ and the absorption coefficient $\alpha_\omega(s)$, are required. In the present model the electrons are assumed to be in thermal equilibrium. This allows the usage of Kirchhoff's law to calculate the emissivity from the absorption coefficient:

$$j_{\omega,n}(s) = \alpha_\omega(s) I_{\mathrm{BB},\omega}(s) \approx \alpha_\omega(s) N^2_{\omega,\,\mathrm{ray}} \frac{\omega^2 T_\mathrm{e} k_\mathrm{b}}{8\pi^3 c_0^2} \tag{7}$$

$I_{\mathrm{BB},\omega}$ is the black body intensity. Applying the Rayleigh-Jeans approximation yields the expression given on the right side of equation (7) for $I_{\mathrm{BB},\omega}$ [11]. Equation (7) can be rearranged to deliver $\frac{j_{\omega,n}(s)}{N^2_{\omega,\mathrm{ray}}}$. Hence, it is enough to compute the ratio $\frac{j_{\omega,n}(s)}{N^2_{\omega,\mathrm{ray}}}$ and this is possible without explicitly evaluating $N^2_{\omega,\,\mathrm{ray}}$, cf. equation (6).

For the absorption coefficient an almost entirely analytic formulation is used, which is given by ref. [12]. The advantage of this method is given by very high numerical stability and the low requirement in computational effort. Such advantages are desirable, if the forward model is to be used in a gradient based optimization framework like IDA (s. a. ref. [10]). For a moderately dense plasma with large T_e the absorption coefficient of the n-th harmonic can be approximated by (s. ref. [12], specifically equation (2a) and (10a)):

$$\alpha_{\omega,\mathrm{n}}(s) \approx 4\pi^2 \frac{{\omega_\mathrm{p}}^2 \mu}{c_0 \omega} \iint \mathrm{d}u_\perp \mathrm{d}u_\parallel \frac{u_\perp}{\gamma} \times$$
$$\underbrace{\left(\frac{n}{\bar{\omega} N_\perp}\right)^2 \left|\left(e_x + \frac{\bar{\omega} N_\perp}{n} u_\parallel e_z\right) J_n(b) - \frac{ib}{n} J_n'(b) e_y\right|^2}_{\text{"Polarization factor"}} \times$$
$$\underbrace{f_{\mathrm{MJ}}(s, u_\perp, u_\parallel)}_{\text{EMD}} \underbrace{\delta\left(\gamma - u_\parallel N_\parallel - \frac{n}{\bar{\omega}}\right)}_{\text{Resonance cond.}} \tag{8}$$

Equation (8) contains an integral in cylindrical, dimensionless momentum space. The symbols $u_{\perp/\parallel} = p_{\perp/\parallel}/(c_0 m_{\mathrm{e},0})$ corresponds to the (dimensionless) momentum perpendicular/parallel to the magnetic field. The Lorentz factor γ is given by $\gamma = \sqrt{1 + u_\perp^2 + u_\parallel^2}$. The inverse normalized cyclotron frequency is given by $\bar{\omega} := \frac{\omega}{\omega_{\mathrm{c},0}}$ and $\mu := \frac{m_{\mathrm{e},0} c_0^2}{T_\mathrm{e}}$. The $\delta(\dots)$ in equation (8) represents a Dirac delta function. The $J_n(b)$ stands for the Bessel function of first kind and n-th order and $b := \bar{\omega} N_\perp u_\perp$. Finally, e_x, e_y, e_z are the components of the polarization vector. The mode selection is performed by the choice of N_ω and the polarization vector. Note that most of these quantities carry an implicit dependence on LOS position s. To allow for arbitrary propagation angles, the quasi-perpendicular approximation (assuming $N_\parallel < v_\mathrm{t}/c_0$ with v_t the thermal electron velocity) that is utilized in ref. [12] is not applied in the implementation. In the forward model equation (8a) of reference [12] is implemented for the calculation of the absorption coefficient. In this equation the individual terms that build up the absorption coefficient are not as clearly visible as in equation (8), but still there are several simplifications that are advantageous for the implementation. The most notably distinction between equation (8) and equation (8a) of reference [12] is that the latter holds the solution of the integral over $u_\perp$ in equation (8). The remaining integration over $u_\parallel$ is performed numerically in the model (Gaussian quadrature). To obtain $\alpha_\omega(s)$ the contribution of all individual harmonics $n = 1, ..., \infty$ has to be summed up. For ASDEX Upgrade typical T_e and for the various ECE viewing geometries available at ASDEX Upgrade [7, 10] it is generally only necessary to consider the contribution of harmonics that have a cold resonance in the plasma. For X-mode ECE diagnostics located on the LFS, the first harmonic is also irrelevant, since the resonance lies in cut-off.

The contributions to the absorption coefficient can be split into three components as shown in equation (8), i.e. the "polarization factor" [12], the electron momentum distribution (EMD) and (relativistic) resonance condition. For the EMD a relativistic Maxwellian is used also known as Maxwell-Juettner distribution. For the meaning and implications of the resonance conditions we refer to ref. [11]. The "polarization factor" accounts not only for wave polarization, but also for finite Larmor radius effects [12]. For the wave polarisation the cold plasma approximation is used in ref. [12] and this is also implemented in the present model. For further details we refer to [2, 12].

2.3 Wall reflections

At frequencies for which the cold resonance lies near the plasma edge or in the scrape-off layer (SOL), the optical depth (τ_ω) is small. Therefore, it is possible that the plasma is not optically thick (i.e. $\tau_\omega < 3$ [2]) after a single pass through the plasma. Since the reflectivity of the wall material is hight the LOS is reflected on the high field side (HFS) and passes through the plasma a second time. Depending on the optical depth of the plasma this process can occur consecutively multiple times, which increases the total length of the LOS and if the reflect LOS passes through resonant layers also the observed intensity. To account for this process, a model of infinite reflections is employed using the wall reflection coefficient R_{wall} as a (free) parameter. Since all ECE diagnostics at ASDEX Upgrade have a slight toroidal inclination the ray does not pass through the LFS port when reflected back from the HFS. Instead the ray hits again a wall tile and is reflected back towards the HFS. This motivates to approximate the vessel wall as two plain mirrors facing each other. Assuming that the radiation of the plasma I_ω and the reflectivity of the wall R_{wall} is constant for each reflection, the relation between observed intensity and the intensity of a single pass is given by [13]:

$$\kappa_{\text{Refl}} := \frac{I_{\omega,\text{Refl}}}{I_\omega} = \frac{1}{1 - R_{\text{wall}} e^{-\tau_\omega}} \tag{9}$$

At low optical depth κ_{Refl} can be larger than two, meaning that the lengthening of the LOS due to wall reflections is a dominant effect. The consequences of this for the validity range of the model is discussed in section 2.3.

2.4 The birthplace distribution of observed intensity

To illustrate which portion of the plasma contributes to the measured intensity, a birthplace distribution of observed intensity (BPD) $D_\omega(s)$ can be defined:

$$D_\omega(s) := \frac{j_\omega(s) T_\omega(s)}{I_\omega(s_{\text{ow}})} \tag{10}$$

In equation (10) the the transmittance $T_\omega(s)$ of the plasma for waves at frequency ω along the LOS from point s to the observation point s_{ow} was introduced:

$$T_\omega(s) := e^{-\tau_\omega(s)} = \exp\left(-\int_s^{s_{\text{ow}}} \alpha_\omega(s')\text{d}s'\right) \tag{11}$$

A function similar to equation (10) is frequently used in the literature (e.g. [14], [15]), but never in conjunction with a proper normalization. To obtain the BPD the integral representation [16] of the radiative transport equation (5) is required, where the boundaries are given by the start and end points of the LOS (e. g. inner wall s_{iw} and outer wall s_{ow}, if there is no internal reflection):

$$I_\omega(s_{\text{ow}}) = \int_{s_{\text{iw}}}^{s_{\text{ow}}} j_\omega(s) T_\omega(s) \text{d}s \tag{12}$$

Normalizing the argument of the integral in equation (12) to the observed intensity $I_\omega(s_{\text{ow}})$ yields the BPD (see equation (10)). How useful the BPD is, is demonstrated using discharge # 30839 at t = 1.75 s as an example. The same channel of the slightly oblique ECE Imaging diagnostic at ASDEX Upgrade is picked as for the benchmark of the ray tracing. In figure 2 the top shows both the emissivity j_ω (blue line) and the transmittance T_ω (black dashed line) as a function of LOS coordinate s. At the bottom the BPD (red line) is shown as well as T_e (dotted black line). One can see that although the emissivity is non-zero for a wide region, the BPD remains zero due to the extremely small transmittance . Even near the cold resonance position (vertical dashed line) the transmittance is still almost zero due to an optical thick layer shortly after the cold resonance. This layer is formed by the electrons, which have their cyclotron frequency up-shifted by the Doppler-shift. In this layer the transmittance then rises to one and the BPD becomes non zero. One can then use the maximum of the BPD as a corrected resonance position or "warm" resonance position [14]. Alternatively one can use the first moment of the BPD as a weighted center of the emission line [17].

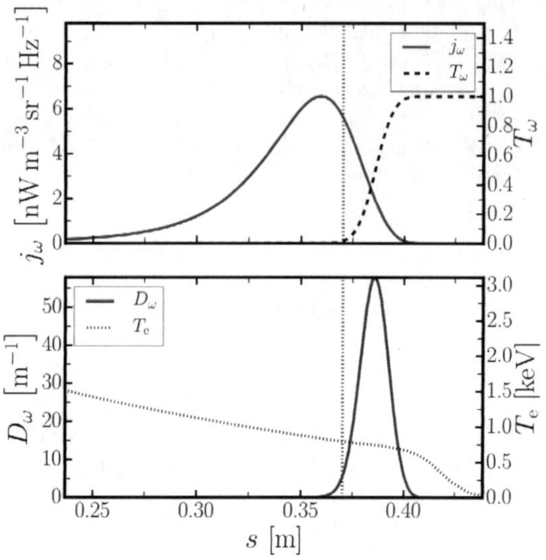

Figure 2. Emissivity, transmittance (top), BPD and T_e (bottom) for # 30839 at t = 1.75 s. The corresponding LOS is shown in figure 1.

3 Limitations of the presented model

The model presented in the previous section has two major limitations. The strongest is the influence of wall reflections or multiple passes through a resonance at extremely low optical depths. The second, weaker limitation are plasmas, where $\omega_c \approx \omega_p$. This section delivers a brief discussion about the validity range of the model.

3.1 Minimum optical depth for accurate modelling

At low enough optical depth wall reflections can enhance the observed T_{rad} by an order of magnitude. Unfortunately accounting for the chaotic ray dynamics originating from multiple reflections of rays on the complex structures of the vessel is computationally far too expensive to be used in forward modelling. Empirically it was observed, however, that the simple model of two plain mirrors facing each other can deliver good results. Using the simplified reflection model presented in section 2.3 for the profile radiometer at ASDEX Upgrade the ECE measurements could be described up to low optical depth [10]. However, the profile radiometer at ASDEX Upgrade is situated in the mid plane. This means that the poloidal inclination of the LOS is only few degrees. Hence, the model of the two mirror represents the situation of the LOS being reflected on the torus wall adequately. For larger poloidal inclinations of the LOS the cold resonance position changes after each reflection since the LOS does not hit the wall perpendicularly. Also in the case of an internal reflections, it is usually the case that the second pass does not have the same cold resonance as the first. Therefore, the assumption that each reflection contributes as much radiation as the first pass is poor. For this reason the impact of wall reflections onto the forward modelled T_{rad} is studied. Given an uncertainty of the measured T_{rad} of 10 %, $\kappa_{\text{Refl}} < 1.1$ for wall reflections to affect the measured T_{rad} also by 10 %. Using equation (9) and a reasonable $R_{\text{wall}} \approx 0.9$ [10] this means that for $\tau_\omega > 2.3$ wall reflections increase the measured T_{rad} by 10 %.

3.2 Limitiations of the analytic absorption formula

The absorption coefficient presented in ref. [12] is optimized to accurately predict the microwave radiation losses through high harmonics in high T_e plasmas. Therefore, the applicability of equation (8) for ECE forward modelling of low needs to be investigated. In the following the approximations used in ref. [12] are reviewed and their effect onto the validity range of the present model are discussed.

In the formulation presented in ref. [12] there are two major approximations that have consequences for the model. The first is that only the electromagnetic component of the energy flux density is considered [12, 15, 18]. The second is that both refractive index and wave polarization are considered in the cold plasma limit. This renders the absorption coefficient to be not applicable at all for the first harmonic [2, 18]. For harmonics with $n > 4$ the harmonic overlap does not allow the expansion into single harmonics [2]. For the regime, where harmonics start to overlap, the sum over infinite harmonics needs to be carried out asymptotically. An analytical formula for this sum can be found in ref. [12]. However, for the application at ASDEX Upgrade the main interest is in the emission of the second harmonic X-mode.

For the second harmonic, strictly speaking, the accuracy of the absorption coefficient is only guaranteed, if [2, 12, 18]:

$$\omega_{\text{c},0}^2 \gg \omega_{\text{p}}^2 \qquad (13)$$

Since the main target of the model is to predict T_{rad} for ECE measurements around the second harmonic X-mode, the applicability of the absorption coefficient given by [12] needs to be discussed in further detail.

Generally the condition given by expression (13) is only met at ASDEX Upgrade for low density plasmas. However, the sensitivity of T_{rad} on the magnitude of the absorption coefficient is low, as long as the optical depth $\tau_\omega > 3$. Only at ECE measured frequencies corresponding to the plasma edge and SOL, where the optical depth is low, T_{rad} becomes dependent on the absolute magnitude of the absorption coefficient. To asses the sensitivity a benchmark was performed using the ASDEX Upgrade profile radiometer antenna pattern (s. [10] for details about this diagnostics). For a wide range of T_e and n_e profiles T_{rad} was then calculated using the model presented in section 2. For the comparison the same scenarios were also modelled using the absorption coefficient derived from the fully relativistic dispersion relation as described by ref. [19]. The results are the following. In general the absorption coefficient used in the present model approximates the fully relativistic absorption coefficient adequately, even if condition given by expression (13) is violated. In most cases the relative deviations between the two absorption coefficients is smaller than 20 %. Even if this threshold is exceeded, the effect on T_{rad} remains small as the absolute value of the absorption coefficient influences the BPD only weakly. However, there is the exception of a resonance coinciding with a reflection. Close to a reflection equation (8) becomes inaccurate, since the condition expressed by equation (13) is strongly violated. Furthermore. the ray tracing is inaccurate due to the omission of relativistic effects. In all other cases the "warm resonances" obtained with both the absorption coefficient presented in section 2 and the one given in ref. [19] deviate only by less than 3 mm as is demonstrated with an example in the following section.

4 Example: Birthplace distribution of observed intensity for the slightly oblique ECE Imaging diagnostic at ASDEX Upgrade

The ECEI at ASDEX Upgrade [7] consists of two separate imaging diagnostics. Through the usage of two opposite signed toroidal angles the two arrays allow measurements in 3D. Although the toroidal deviation from the perpendicular is small (e.g. about 7° at launch for #30839), the Doppler-Effect has significant influence onto the center of the BPD. This is demonstrated using the following example. In discharge # 30839 the ECEI system was used to measure the plasma displacement by magnetic perturbation coils [20]. For this purpose the ECEI observed the plasma edge near the separatrix. As described in ref. [20] the measurement position is not adequately captured by the cold resonances. However, through the usage of the model described in section 2 it was possible to recover the actual measurement positions using the maximum of the BPD as "warm" resonance position.

In figure 3 the expected measurement positions of the first ECEI array are shown in the cross section of ASDEX Upgrade for discharge #30839 at $t = 1.75$ s. All channels for which the measured optical depth $\tau < 0.05$ were removed from the plot, since they lie far outside the validity regime of the model. If one assumes that the measurement positions are well described by the cold resonances (red stars) then, as shown in figure 3, one would expect the array to cover a large portion of the plasma edge and

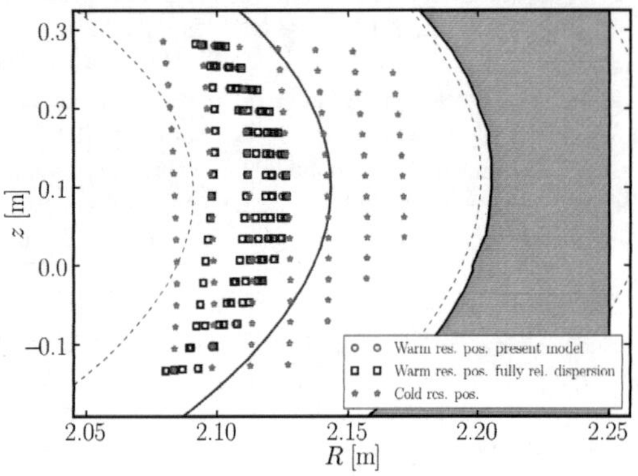

Figure 3. Poloidal cross section of ASDEX Upgrade for discharge # 30839 at t = 1.75 s. The dashed black lines show flux surfaces and the blue line marks the separatrix. The grey structure outlined by a solid black line depicts a piece of the outer vessel wall.

the SOL. But, if radiation transport is considered, the expected measurement position shrink together to a thin band shortly inside the separatrix (green circles). For the channels in the SOL this is due to the shine-through originating from the pedestal top. For the channels further inside the plasma the "warm" resonances are shifted outward by the Doppler effect. In figure 3 a third set of warm resonance positions, marked by blue squares, is shown. These warm resonances are obtained if the absorption coefficient discussed in section 2 is replaced by the absorption coefficient obtained from the fully relativistic dispersion relation (s. ref [19]). The "warm" resonance positions according to the absorption coefficient presented in section 2 and the absorption coefficient obtained from the fully relativistic dispersion relation are displaced at their maximum by 2 mm. This is much smaller than the accuracy of the measurements used for the prediction of the model. Hence, the absorption coefficient presented in section 2 is at least sufficient in this case for the purposes of determining warm resonances for oblique ECE diagnostics like the ECEI diagnostic at ASDEX Upgrade.

5 Conclusions

A model for the radiation transport consideration refraction in the cold plasma limit has been presented. This model makes use of the approximation for the absorption coefficient presented in ref. [12]. The limitations of the model due wall reflections and internal reflections at low optical depth have been discussed. Furthermore it was shown that for the slightly oblique ECEI at ASDEX Upgrade the "warm" resonances obtained can differ significantly from the cold resonances. Lastly it was demonstrated, that the approximation used for the absorption coefficient is adequate for the calculation of warm resonances for oblique ECE diagnostics.

Acknowledgements

The author acknowledges Daniela Farina and Lorenzo Figini for providing a code for the calculation of the absorption coefficient from the fully relativistic dispersion relation.

This work has been carried out within the framework of the EUROfusion Consortium and has received funding from the Euratom research and training programme 2014-2018 under grant agreement No 633053. The views and opinions expressed herein do not necessarily reflect those of the European Commission.

References

[1] D. Farina, FUSION SCI TECHNOL **52**, 154 (2007)
[2] M. Bornatici et al., Nuclear Fusion **23**, 1153 (1983)
[3] R. Dux et al., *Tungsten as first wall material in ASDEX Upgrade*, in *21st IAEA Fusion Energy Conference* (International Atomic Energy Agency, 2007)
[4] A.C. Hindmarsh, Scientific Computing **1**, 55 (1983)
[5] P. Dierckx, *Curve and surface fitting with splines* (Oxford University Press, 1995)
[6] E. Poli et al., Comput. Phys. Commun. **136**, 90 (2001)
[7] I.G.J. Classen et al., Rev. Sci. Instrum **85**, (2014)
[8] R. Prater et al., Nuclear Fusion **48**, 035006 (2008)
[9] G. Bekefi, *Radiation processes in plasmas*, Wiley series in plasma physics (Wiley, 1966)
[10] S.K. Rathgeber et al., PPCF **55**, 025004 (2013)
[11] I. Hutchinson, *Principles of Plasma Diagnostics* (Cambridge University Press, 1987)
[12] F. Albajar et al., PPCF **49**, 15 (2007)
[13] N. Luhmann Jr et al., Fusion Sci. Technol. **53**, 335 (2008)
[14] G. Garstka, M. Austin, R. Ellis, Fusion Engineering and Design **53**, 123 (2001)
[15] A.I. Smirnov et al., JETP **83**, 294 (1996)
[16] L. Yong et al., Chinese Physics B **21**, 045201 (2012)
[17] N.B. Marushchenko et al., FUSION SCI TECHNOL **50**, 395 (2006)
[18] M. Bornatici et al., PHYS PLASMAS **1**, 189 (1994)
[19] D. Farina, FUSION SCI TECHNOL **53**, 130 (2008)
[20] M. Willensdorfer et al., submitted to PPCF (2016)

Update on the status of the ITER ECE diagnostic design

G. Taylor[1a], M. E. Austin[2], A. Basile[1], J. H. Beno[3], S. Danani[4], R. Feder[1], S. Houshmandyar[2], A. E. Hubbard[5], D. W. Johnson[1], A. Khodak[1], R. Kumar[4], S. Kumar[4], A. Ouroua[3], S. B. Padasalagi[4], H. K. B. Pandya[4], P. E. Phillips[2], W. L. Rowan[2], J. Stillerman[5], S. Thomas[4], V. S. Udintsev[6], G. Vayakis [6], M. Walsh[6], and D. Weeks[3]

[1]Princeton Plasma Physics Laboratory, Princeton, NJ 08543, USA
[2]Institute for Fusion Studies, University of Texas at Austin, TX 78712, USA
[3]Center for Electromechanics, University of Texas at Austin, TX 78758, USA
[4]ITER-India/Institute for Plasma Research, Bhat 382428, Gandhinagar, India
[5]Plasma Science and Fusion Center, MIT, Cambridge, MA 02139, USA
[6]ITER Organization, Route de Vinon sur Verdon, 13115, St Paul Lez Durance, France

Abstract. Considerable progress has been made on the design of the ITER electron cyclotron emission (ECE) diagnostic over the past two years. Radial and oblique views are still included in the design in order to measure distortions in the electron momentum distribution, but the oblique view has been redirected to reduce stray millimeter radiation from the electron cyclotron heating system. A major challenge has been designing the 1000 K calibration sources and remotely activated mirrors located in the ECE diagnostic shield module (DSM) in the equatorial port plug #09. These critical systems are being modeled and prototypes are being developed. Providing adequate neutron shielding in the DSM while allowing sufficient space for optical components is also a significant challenge. Four 45-meter long low-loss transmission lines transport the 70-1000 GHz ECE from the DSM to the ECE instrumentation room. Prototype transmission lines are being tested, as are the polarization splitter modules that separate O-mode and X-mode polarized ECE. A highly integrated prototype 200-300 GHz radiometer is being tested on the DIII-D tokamak in the USA. Design activities also include integration of ECE signals into the ITER plasma control system and determining the hardware and software architecture needed to control and calibrate the ECE instruments.

1 Introduction

Considerable progress has been made on the design of the ITER electron cyclotron emission (ECE) diagnostic since the status of the design was presented at the EC-18 workshop in Nara, Japan in April 2014 [1]. The ECE diagnostic will provide important information on the time evolution of the electron temperature profile ($T_e(R)$), magneto-hydrodynamic (MHD) fluctuation spectra, non-thermal electron behavior and the ECE radiated power loss. The design is being carried out through a close collaboration between teams in the USA domestic agency (US-DA), India Domestic Agency (IN-DA) and at the ITER Organization (IO) in France. The design is currently midway through the detailed preliminary design phase that will culminate in a preliminary design review in 2017. A radial and oblique view [2] are still included in the design, but since 2014 the oblique view has been redirected to reduce collection of stray millimeter radiation from the electron cyclotron heating (ECH) system. A major challenge has been designing the front-end ECE diagnostic shield module (DSM) in equatorial port plug #09, and the components located therein. These components include two 1000 K calibration sources and remotely activated mirrors that switch the ECE instruments from viewing the plasma to viewing the in-situ calibration sources in the DSM. These components must operate reliably over the 20-30 year lifetime of ITER, they are being modeled and prototypes are being developed in order to guide the design process. In particular, providing adequate neutron shielding in the DSM while allowing sufficient space for the calibration sources and optical components is a major challenge. Another design challenge is developing the four 45-meter long low-loss transmission lines that transport the 70-1000 GHz ECE from the DSM to the ECE instrumentation room in the ITER Diagnostics Hall. Prototype transmission lines are being tested at the IN-DA, as is the polarization splitter module that will separate the O-mode and X-mode polarized ECE at the front of the transmission lines. The ECE diagnostic is required to provide real-time signals to help guide the deposition location for ECH to suppress neoclassical tearing modes (NTMs) and other MHD activity, and for $T_e(R)$ control. As part of the preliminary design activities,

[a] Corresponding author: gtaylor@pppl.gov

integration of the ECE signals into the plasma control system (PCS) is being addressed. This work includes determining the hardware and software architecture needed to control and calibrate the ECE instruments, and to store and analyze the ECE data. This paper updates the status of these design and prototyping activities. Section 2 covers the design of front end components, section 3 covers the design of components between the DSM and the ECE instrumentation room, and section 4 describes instrumentation in the ECE room, and provides an overview of the software needs and information flow between ECE instrumentation and the ITER plant, including the PCS.

2 Front-end components

The DSM design has significantly changed over the past two years in order to address issues related to neutron shielding and structural integrity. There have also been

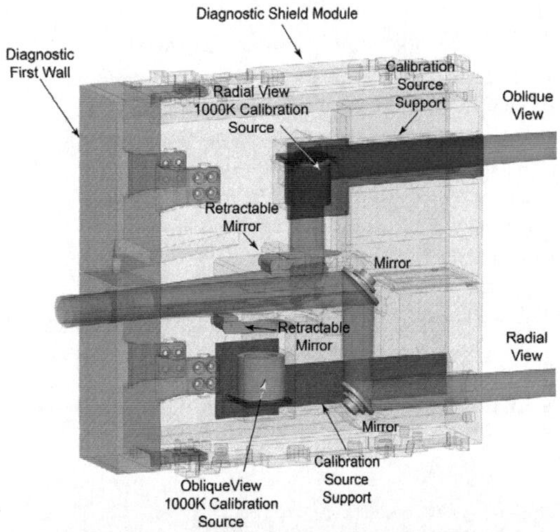

Figure 1. Side view of front-end optics in the ECE DSM showing calibration sources and retractable mirrors that switch in the calibration sources to the radial and oblique views and calibration source support brackets (blue).

changes in the optical design and in how the mirrors that switch between the plasma views and the calibration sources are controlled. Figure 1 shows a side view of the ECE DSM. The DSM is now comprised of upper and lower shielding modules that include internal spaces for optical components. Previously the DSM had three shielding sections split vertically but this complicated connections for water cooling. It would be even better to have just one shielding module but this cannot be manufactured using conventional machining. It is concievable in the future that such a single shielding module could be created with a 3D printing additive manufacturing technique. The diagnostic first wall (DFW) module is bolted to the two shielding modules at the front of the DSM. The concept for supporting the calibration sources is presently to bolt them to plates intalled on the outside of the DSM (Fig. 2). These plates extend to the back of the DSM and include a channel machined inside to carry cables.

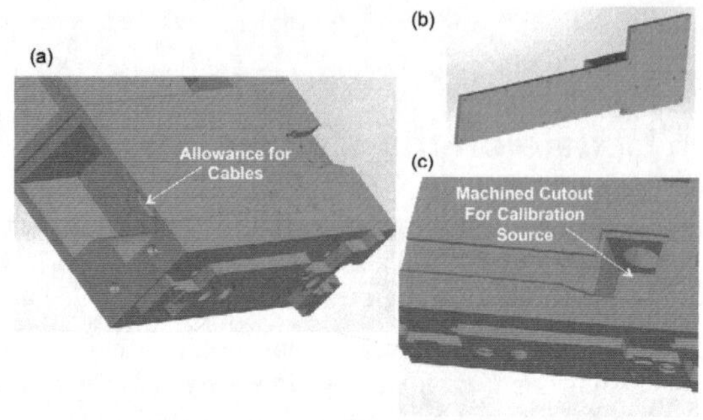

Figure 2. Calibration source support plate concept. (a) Rear view of ECE DSM, (b) hot source mounting plate, and (c) bottom view of ECE DSM.

The width of the DSM was reduced in 2014 requiring the oblique viewing angle to be reduced from 13° to 11.5° from radial (Fig. 3). Also the oblique view direction was redirected to the opposite toroidal direction to reduce the possibility of stray millimeter wave radiation entering from the ECH system.

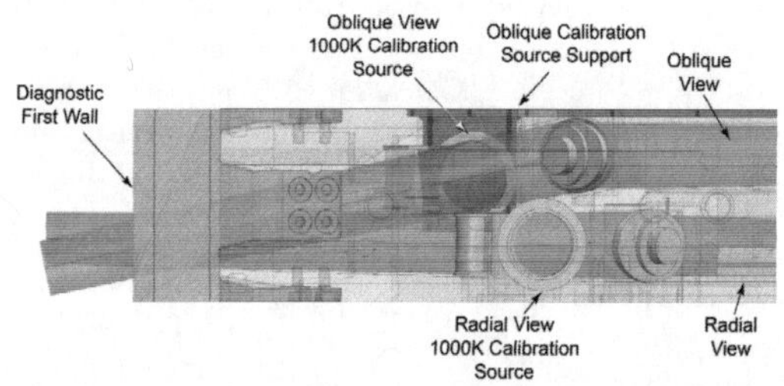

Figure 3. Top view of ECE DSM showing calibration source locations.

During the past two years there has been significant design and prototyping activity in the USA on the hot calibration sources that will be located in the DSM. A prototype design (Fig. 4) is being developed that minimizes the use of brittle materials to mitigate high shock loads, avoids direct water cooling to reduce failure points, and uses indirect cooling from the DSM wall. The design uses a Nichrome heating element embedded in an Inconel heating block for improved mechanical support and protection. Two heating configurations are being explored; indirect heating of the Silicon Carbide (SiC) emitter through radiation to avoid direct mechanical contact between the metallic surface and the ceramic emitter back surface, and direct contact heating where the heater is in contact with the emitter. While indirect heating is the preferred option, since it has mechanical benefits, it was found in early in-vacuum testing of a commercial MeiVac [3] embedded Inconel heater that surface emissivity not only varies significantly with temperature, but also due to the formation and sublimation of oxide layers during heating. A vibration

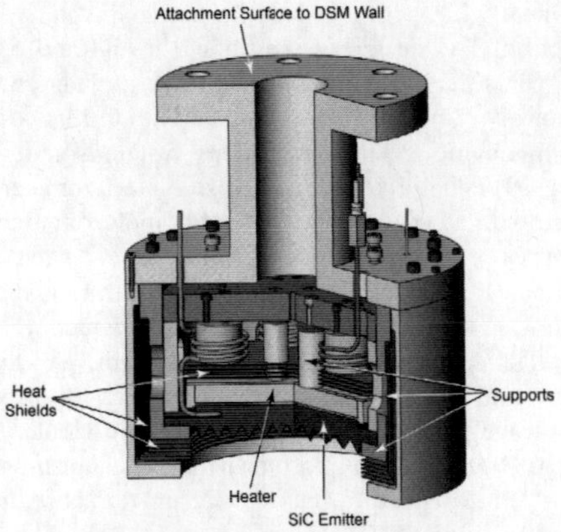

Figure 4. Calibration source showing the direct-contact heating option.

analysis of the ECE hot sources is being conducted. There are various fast transient electromagnetic loads that can result in vibrations of the calibrations sources. These electromagnetic loads can be driven by major disruptions and plasma vertical displacement events that induce inertial loads on the Port Plug and DSM. Results of the vibration analysis so far indicate these loads can induce 50-60 Hz vibrations with a maximum acceleration of 8 g.

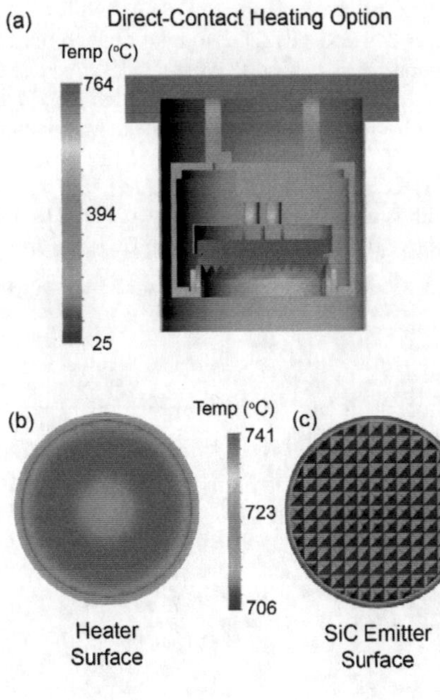

Figure 5. Thermal analysis of a hot source using the direct-contact heating and 1.2 kW of heating power. (a) a vertical section through the calibration source. (b) The heater surface and (c) the SiC emitter surface. The color scales indicate the predicted temperatures of source components.

Detailed thermal analysis (Fig. 5) and experimental evaluation of prototype hot sources with direct and indirect heating of the SiC emitter is currently being conducted. So far good agreement has been found between the thermal analysis and experiment.

The design presented at EC-18 had wire rope actuators controlling the remotely activated mirrors that switch between viewing the calibration sources and the plasma. The wire rope concept required four penetrations through the port plug back plate that increased neutron streaming and also required sophisticated remote handling. One concept being considered now is to use piezoelectric actuators mounted on the mirrors. These actuators can be made magnetic field and ultra-high vacuum compatible, and require no mechanical feedthroughs. The piezoelectric actuators can be installed on the remotely actuated mirrors in such a way as to allow the mirrors to act as shutters blocking the calibration sources when they are retracted for plasma ECE measurements, this was not possible with the wire rope design. Furthermore commercial piezoelectric actuators are available that have two piezoelectric drives on the same axle. This arrangement provides redundancy if one of the piezoelectric drives were to fail. Available commercial piezoelectric actuators have a pointing accuracy of < 0.1° and can operate at up to 200 °C. Prototype testing of a piezoelectric actuator will be conducted during 2016. If the prototype testing shows the concept is viable, the piezoelectric actuator will be subjected to a neutron fluence corresponding to the lifetime exposure it would be subjected to in the ECE DSM. Another concept being considered is a single rod actuator for each remotely activated mirror. The mirror would be held in the plasma viewing position by a sprirng or counter weight and the rod would push on the mirror to switch the mirror to view the calibration source.

3 Components between the front-end and the instrumentation room

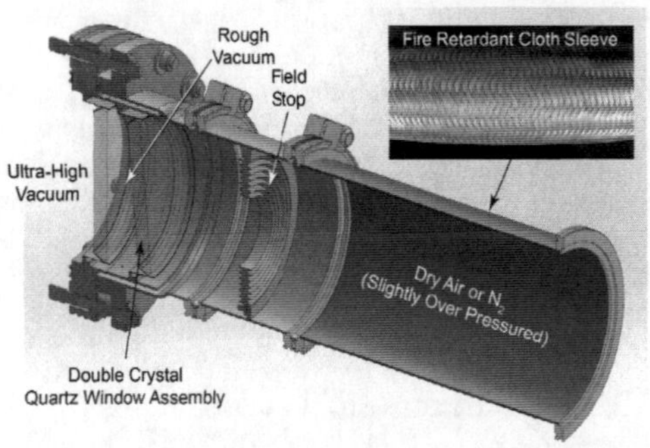

Figure 6. Concept for the interface between the vacuum window assembly and the polarization splitters.

The vacuum window assemblies at the back of the DSM will be provided by ITER but the design of the interface between the vacuum windows and the polarization splitters is the responsibility of the US-DA. The remainder of the transmission line system connecting the front-end to the ECE room in the Diagnostics Hall, including the polarization splitters, is the responsibility of the IN-DA.

Figure 6 shows the present concept for the interface between the vacuum window assembly and the polarization splitters. A fire retardant cloth sleeve that is slightly over pressured with either dry air or nitrogen provides a compliant coupling between the vacuum window assembly and the polarization splitters. Controlled leakage of over pressured gas can ensure evacuation of water vapor from the transmission line. A field stop, probably made from SiC, will be located between the double vacuum window assembly [4] and the compliant coupling to filter unwanted high order modes.

A prototype polarization splitter box is being assembled by the IN-DA. Each polarization splitter box contains two Gaussian telescopes constructed from three

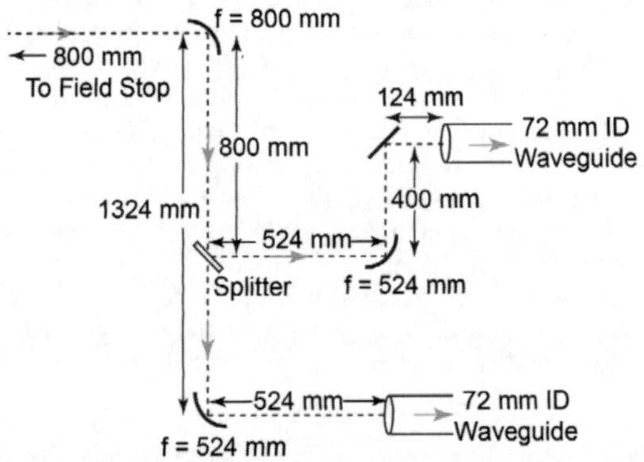

Figure 7. Schematic diagram of the polarization splitter.

ellipsoidal mirrors (Fig. 7). The calculated cross polarization loss is 31.1 dB at the 800 mm focal length mirror and 30.4 dB at the two 524 mm focal length mirrors. The calculated higher order mode loss is 34.1 dB at the 800 mm focal length mirror and 33.4 dB at the two 524 mm focal length mirrors. 10-15 m lengths of prototype smooth walled circular waveguide transmission line sections with an internal diameter (ID) of 72 mm have also been fabricated for testing the attenuation. The transmission lines will be evacuated to mitigate absorption by water vapor.

4 ECE instrumentation room

Two, high throughput, reciprocating Martin-Puplett Fourier Transform Spectrometers with a spectral range of 70-1000 GHz, a frequency resolution ≤ 5 GHz and a scanning repetition rate ≤ 20 ms will be provided by the IN-DA. The IN-DA has ordered a prototype Michelson interferometer from Blue Sky Spectroscopy [5]. The instrument will operate in vacuum to avoid water vapour absorption. The design of the optical layout and optical components has been completed and a dual channel detector system has been tested. The testing of the scanning engine and laser metrology system is underway. During 2017 the prototype Michelson interferometer will be used with a 500°C blackbody calibration source to test prototypes of the transmission line waveguide components, including the prototype polarization splitter. The IN-DA will also provide the "Low Frequency" 122-230 GHz heterodyne radiometer system. A hybrid splitter, consisting of quasi-optical diplexer followed by two waveguide diplexers, will distribute the 122-230 GHz radiation into four bands. The quasi-optical splitter unit will use a Gaussian beam telescope and a frequency selective Dichroic beam splitter. The four receivers will have a total of 58 channels: a 16 channel 122-138 GHz receiver with 0.5 GHz bandwidth per channel, a 14 channel 141-168 GHz receiver, a 15 channel 172-200 GHz receiver and 13 channel 205-230 GHz receiver. All three higher frequency receivers will have 1 GHz bandwidth per channel. The IN-DA is developing a high temperature calibration source (6) that will be used in the ECE instrumentation room to allow more frequent calibration of the ECE instruments than will be possible using the calibration sources in the DSM.

The US-DA is responsible for providing the "High Frequency" radiometer system. Originally this system was to cover 244-355 GHz, but to better match the ITER plasma operating scenarios the frequency coverage was changed to 220-340 GHz and the number of standard resolution (1 GHz bandwidth) channels was increased from 48 to 60. There will also be 16 high-resolution (200-250 MHz bandwidth) channels that use YIG filters to tune within a 2-18 GHz frequency range. The mixer bank will have three 2-40 GHz filters. There will be two,

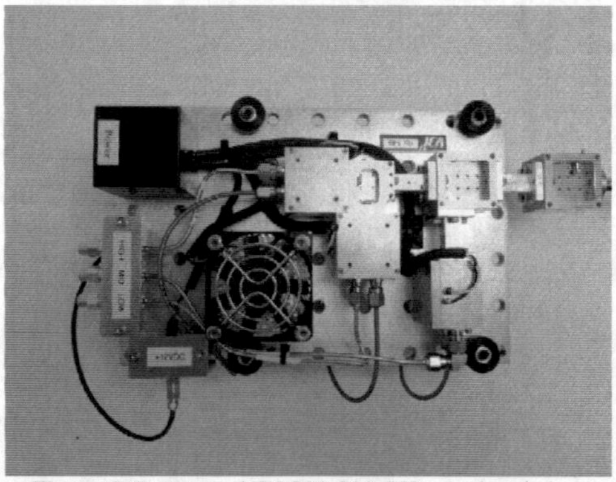

Figure 8. Prototype VDI 200-300 GHz receiver being tested at DIII-D.

temperature-controlled, 30 dB intermediate frequency (IF) amplifiers per bank, with adjustable attenuators covering 0-30 dB in 1 dB steps. Pad attenuators in front of each detector will be used to fine tune sensitivity. The video amplifiers will have a voltage gain of 100-400 and a bandwidth of 500 kHz. An important overall goal of the design for the IF filter bank is to achieve stable response and good sensitivity for each channel and easy access for adjustments and changes. A prototype 200-300 GHz radiometer (Fig. 8), that employs highly-integrated millimeter wave technology, developed by Virginia Diodes [7], is being tested on the DIII-D tokamak at General Atomics in the USA [8, 9]. Third and fourth harmonic plasma ECE measurements from the prototype radiometer are being compared to data from a Michelson interferometer.

Because there are only two Michelson interferometers in the baseline ECE instrumentation there will need to be switches in the ECE room that switch the inputs of the two Michelson interferometers between the radial O-mode, the radial X-mode, the oblique O-mode and the oblique X-mode transmission lines in order to check that the ECE spectrum is thermal and to measure the total ECE power. An ad hoc working group with members from the IN-DA, US-DA and IO was setup to identify the simplest way to achieve this flexibility and agree on the layout of instrumentation in the ECE room.

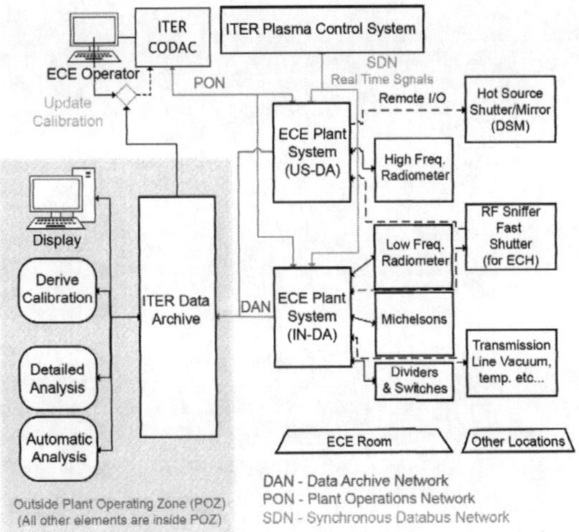

Figure 9. Schematic diagram of information flow for major ITER ECE system elements. Data to and from the ECE Plant Systems will be via the Plant Operations Nework (PON, blue), the Synchronous Databus Network (SDN, green), and the Data Archive Network (DAN, purple)

Another important aspect of the preliminary design phase is to decide the software needed to control and calibrate the ECE instrumentation, and to record and analyze the ECE data. Unlike current fusion experiments, ITER employs a concept referred to as the "Plant Operating Zone" (POZ). Because ITER is a nuclear facility the POZ requires strict safeguards. Figure 9 shows a simplified schematic of the ECE systems and data networks within and outside the POZ. There are two ECE Plant Systems that send and receive data inside the POZ via a relatively slow Plant Operations Network (PON). Signals to the PCS for NTM and $T_e(R)$ control will be sent over a faster Synchronous Databus Network (SDN). Another ad hoc working group with members from the IN-DA, US-DA and IO has been setup to determine what real time signals need to be provided for NTM and $T_e(R)$ profile control. For $T_e(R)$ control 32 channels of $T_e(R)$ data, with channels separated by a/30 (where a is the plasma minor radius) at a digital sampling rate of 100 Hz and with a latency of 2.5 ms should be sufficient, but for NTM control the data requirements are more challenging. An n = 2 NTM may have frequencies up to 10 kHz, requiring digital data sampling rates up to 30 kHz. However the rate of change of NTM parameters would be much slower (~ 1 ms). Radial coverage of the $T_e(R)$ data for NTM control will need to be closely spaced with a channel spacing of a/80 over the region between q=1 and q=2. Options for providing T_e data to the PCS are under consideration by the ECE and ITER PCS teams. The ECE diagnostic system could supply a set of $T_e(t)$ signals at a rate greater than the NTM frequency and/or use high level data processing in the vicinity of the ECE diagnostics to supply NTM parameters at a lower rate of ~1 kHz. These and other details will be determined as the control approaches are decided and the ECE instrumentation and control system designs progress.

5 Summary

Through close collaboration, design teams in India, the US and at the ITER site in France have made significant progress during the past two years designing the ITER ECE diagnostic system, despite the significant technical challenges presented by the ITER environment. In order to guide the design process prototypes of key system components, including a 200-300 GHz radiometer, a 70-1000 GHz Michelson interferometer, a 1000 K hot source, piezoelectric actuator for the shutter/mirrors in the DSM, and broadband quasi-optical transmission line components are being fabricated and tested. Prototype testing will help guide the preliminary design process that will culminate in a Preliminary Design Review in 2017.

Acknowledgements

This work is supported by US DOE Contract No. DE-AC02-09CH11466. All US activities are managed by the US ITER Project Office, hosted by Oak Ridge National Laboratory with partner laboratories located at the Princeton Plasma Physics Laboratory and the Savannah River National Laboratory. The project is being accomplished through a collaboration of DOE Laboratories, universities and industry. The views and opinions expressed herein do not necessarily reflect those of the ITER Organization.

References

1. G. Taylor, *et al.*, EPJ Web of Conferences **87**, 03002 (2015); DOI: 10.1051/epjconf/20158703002.
2. G. Taylor and R. W. Harvey, Fus. Sci, Tech. **55**, 64 (2009).
3. MeiVac Inc heater : http://www.meivac.com/htr-substrate-heater-family/

4. V. S. Udinstev, *et al*., "Primary window assemblies for microwave diagnostics in ITER", Paper TECH-P-4, EC-19 Conference (2016).
5. Blue Sky Spectroscopy: http://blueskyspectroscopy.com/?page_id=13
6. R. Kumar, *et al*., "Characterization of in-house developed high temperature black body source in frequency range 70-1000 GHz", Paper ECE-P-5, EC-19 Conference (2016).
7. E. W. Bryerton, *et al*, "A compact and reliable 200-300 GHz receiver for the ITER ECE system", Proc. 39th Int. Conf. on Infrared, Millimeter, and Terahertz Waves, Tucson, AZ, USA (2014); DOI: 10.1109/IRMMW-THz.2014.6956470
8. M. E. Austin, *et al.,* "Testing of the ITER-ECE prototype receiver and related components on DIII-D", Paper JP12.00070 Bull. Am. Phys. Soc. **60** No. 19. (2015).
9. S. Danani, *et al*., "Testing of the Prototype Receiver for ITER ECE Diagnostic", Paper ECE-O-3, EC-19 Conference (2016).

EPJ Web of Conferences **147**, 02004 (2017) DOI: 10.1051/epjconf/201714702004
EC-19

Testing of the Prototype Receiver for ITER ECE Diagnostic

S. Danani[1,*], *M. E.* Austin[2], *M.W.* Brookman[2], *H. K. B.* Pandya[1], and *Vinay* Kumar[1]

[1]ITER-India, Institute for Plasma Research, Bhat, Gandhinagar-382428, India
[2]University of Texas at Austin, Austin, TX, USA

Abstract. The Electron Cyclotron Emission (ECE) diagnostic system in ITER provides essential information for plasma control and for evaluating the plasma performance. It measures the electron temperature profile (edge/core), electron temperature fluctuations and radiated power in the electron cyclotron frequency range from the plasma. Virginia Diodes Inc. (VDI) has designed and developed a state-of-the-art prototype receiver to detect emission in the 200-300 GHz frequency range. The prototype receiver developed by VDI has been tested at DIII-D tokamak to check its performance by measuring higher frequency ECE harmonics in high temperature plasmas. Bench testing verifies the receiver's band coverage and noise temperatures. Good sensitivity has been obtained meeting the ITER requirement. Also, the receiver performance is assessed for ITER by comparing the data from the receiver and the existing Michelson interferometer for the same DIII-D plasma shot. The measurement results show that the receiver provides accurate temperature information for the plasma emission throughout the frequency band.

1 Introduction

For ITER high toroidal field operation (B_{t0} ~5.3 T), the first and second harmonic ECE extends from 122-230 GHz, and 220-340 GHz respectively. Thus the ITER ECE receivers need to be extremely broadband (~100 GHz) for instantaneous frequency coverage and should have very good sensitivity, linearity across the very wide band, and very high reliability [1]. Similar measurement instruments have been successfully deployed in some tokamaks, like the high frequency radiometer installed in Alcator-C which operates from 234-306 GHz [2]. But the ITER system presents significant new challenges in terms of reliability and robustness. To achieve these goals, Virginia Diodes Inc. (VDI) has designed and developed prototype receiver based on advanced solid state technology [3]. The prototype receiver developed by VDI for frequency range 200-300 GHz was installed at DIII-D to check its performance by carrying out some bench-top measurements, and then measurement of ECE radiation temperature of the DIII-D plasma to test and verify other important parameters [4]. The DIII-D ECE frequency range for third to fifth harmonics that extends from 120-300 GHz and the typical plasma electron temperatures of order 5 KeV, makes it suitable for testing of ITER ECE prototype receiver. In this paper, the test results of noise temperature and sensitivity of the prototype receiver are presented. The receiver performance is assessed for ITER requirement by comparison of ECE measurements with those from the DIII-D Michelson interferometer.

2 Prototype Receiver

The prototype receiver from VDI (Rx149), is a compact solid state millimeter wave receiver system which employs a waveguide-based triplexer, a Dielectric Resonator Oscillator (DRO) based local oscillator, an integrated tripler, subharmonic mixer and amplifier. This integrated assembly leads to improved performance and reliability. The waveguide based triplexer divides the frequency range 200-300 GHz into three bands, low, mid, and high, which then go to respective mixer modules as shown in Fig. 1. These mixers are driven by DROs whose frequencies are set such that each frequency band is down-converted to identical IF band 5-38.33 GHz.

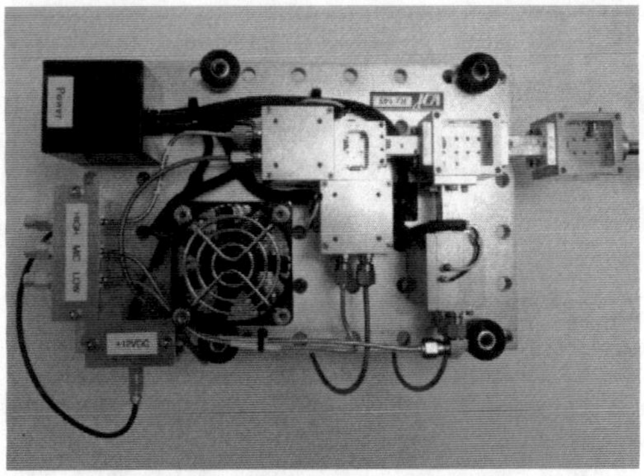

Fig. 1. VDI Prototype Receiver

This satisfies another important requirement for ITER radiometer to have identical IF section in order to

* Corresponding author: suman.danani@iter-india.org

DOI: 10.1051/epjconf/201714702004

have a simple and cost-effective system. The frequency range corresponding to the low, mid and high bands is 200-233.33 GHz, 233.33-266.66 GHz, and 266.66-300 GHz respectively.

3 Bench Testing of the Receiver

The receiver has been first tested with the VDI AMC 446 source. This is an amplifier-multiplier chain with input frequency 11-17 GHz, multiplication factor of 18, output frequency range 200-300 GHz and output power of 400 μW. This measurement is carried out to test receiver's band coverage for each of the three IF bands (high, mid, and low) and is shown in Fig. 2.

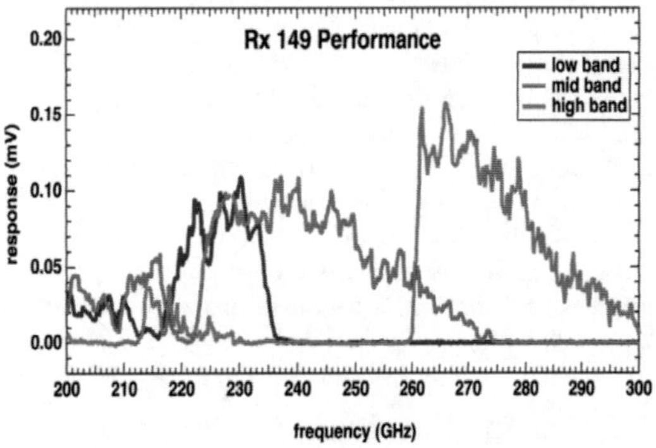

Fig. 2. Receiver performance for the three bands

3.1 Noise Temperature Measurement

An experiment was carried out to measure the receiver noise temperature. A W-band noise tube (CLARE TN-165) was connected to the receiver input via a WR-10 to WR-3.5 waveguide transition. This noise tube radiates at 8000 K in the W-band. First, the noise tube spectrum was obtained using Michelson interferometer and then the noise tube radiation temperature in frequency range 50 to 400 GHz was obtained by comparing the noise tube spectrum with the LN blackbody source spectrum as shown in Fig. 3.

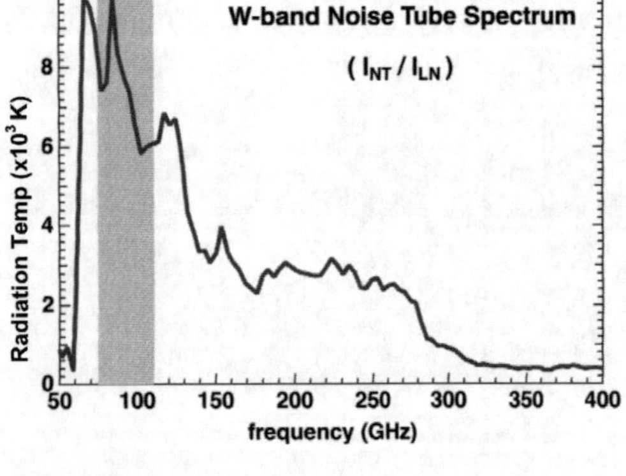

Fig. 3. Noise Tube Spectrum

Measurement indicates that the radiation temperature of the tube is at 3000 K for the desired receiver frequency range.

The receiver noise temperature was obtained using the Y-factor method [5],

$$T_r = (T_h - YT_c) / (Y-1) \tag{1}$$

with T_h = 3000 K and T_c = 300 K. The measurement of noise temperature for the three IF bands for a set of five cavity filters with centre frequencies between 2-18 GHz is shown in Fig. 4. The values obtained are in reasonable agreement with the theoretically estimated value from VDI of ~12000 K, for mixer noise temperature of ~2900 K.

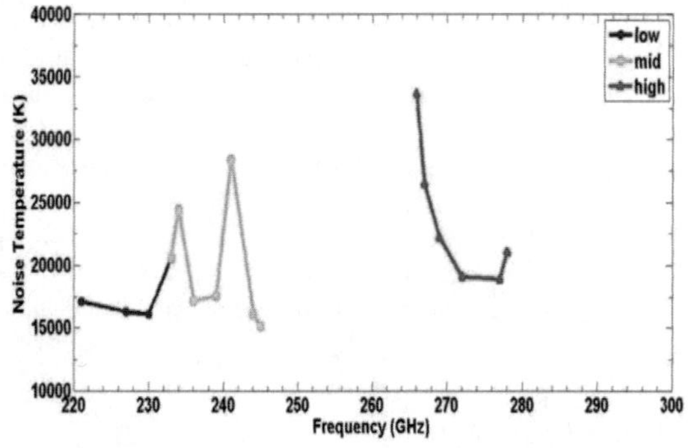

Fig. 4. Receiver Noise Temperature

3.2 Sensitivity measurement

VDI AMC 446 source was also used for the sensitivity measurement. First, output power from this source was measured for the full frequency band (200-300 GHz) using a power meter so that the input power at the receiver is known. This is shown in Fig. 5. For sensitivity measurement the output power was measured for the same set of filters for the three bands. Using the data of Fig. 5, an average sensitivity is obtained of ~ 500 mV/mW and is shown in Fig. 6.

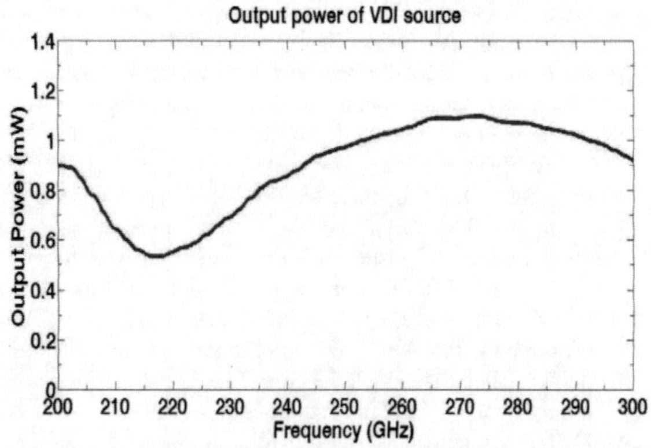

Fig. 5. Output power of VDI AMC 446 source

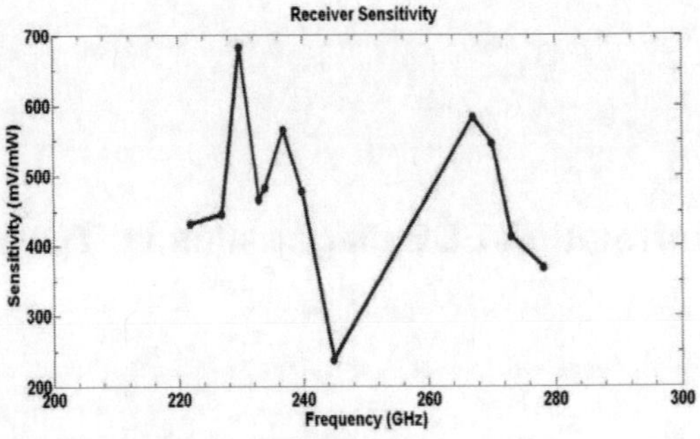

Fig. 6. Plot of receiver sensitivity

4 Receiver Testing On DIII-D

The prototype receiver was installed at DIII-D in a shared optical path with the Michelson interferometer [4] to facilitate simultaneous ECE measurement from both the instruments. The wire grid beam splitter positioned at 45° to the horizontal divides the beam into two: one is coupled to the Michelson and, other to the receiver as shown in Fig. 7.

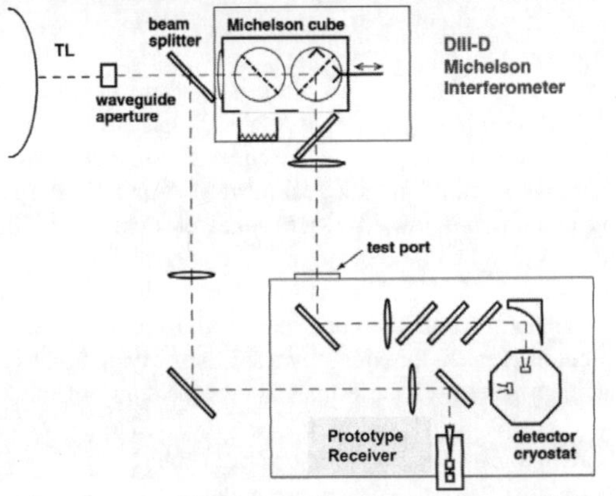

Fig. 7. Schematic of receiver set-up with Michelson at DIII-D

Fig. 8 shows the prototype receiver installed at DIII-D on the Michelson optics table.

Fig. 8. Prototype receiver installed on DIII-D

A novel high intensity (3000 lumens) LED lamp source was used for the alignment of coupling optics of the receiver. The source along with an aperture is placed at the waveguide end. The image is mapped through the coupling optics into a well-defined beam spot at the opening of the receiver input horn.

4.1 Plasma ECE Measurement

Fig.9 shows T_{rad} from prototype receiver and Michelson interferometer for DIII-D shot # 166171. It is observed that both measurements are well matched indicating that the receiver is measuring the proper frequency band. Also its response is linear.

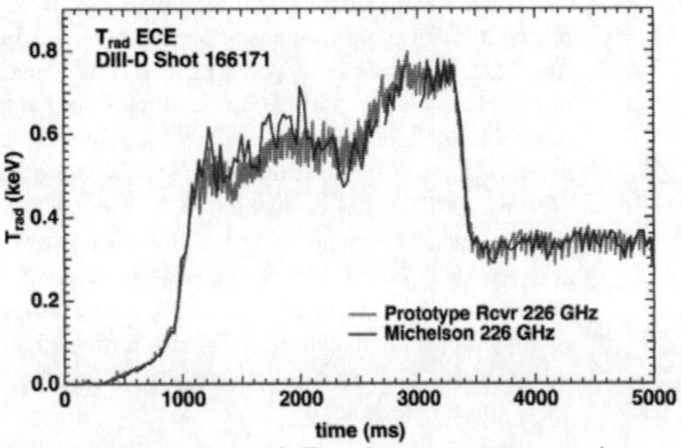

Fig. 9. Comparison of T_{rad} from prototype receiver and Michelson interferometer for DIII-D shot 166171

5 Summary

Prototype receiver for ITER ECE developed by VDI has been installed at DIII-D for high harmonic ECE measurements. Bench testing of prototype receiver show reasonable values for noise temperature, sensitivity, and frequency band coverage. Measurements of DIII-D high harmonic ECE show that the receiver has linear response. Test results shows that the receiver meets ITER requirements in terms of noise temperature, sensitivity, linearity, and frequency coverage. The steady performance of the receiver even after continuous use for a long period proves its robustness.

References

1. G. Taylor et al, PPPL-5026, May 2014
2. R. Chatterjee, P.E. Phillips, J. Heard, C. Watts, R. Gandy, A. Hubbard, Fusion Eng Des **53** 113(2001)
3. E.W. Bryerton, D. Koller, J. L. Hesler, and T.W. Crowe, 39th Int. Conf. on Infrared, Millimeter, and THz Waves, Tucson, AZ, September 12-19, (2014)
4. M. E. Austin, M.W. Brookman, P.E. Phillips, W.L. Rowan, S. Danani, Poster Presented at APS/DPP, Savannah, GA (2015)
5. H. J. Hartfuss and Thomas Geist, *"Fusion Plasma Diagnostics with mm waves: An Introduction"* Wiley Publications (2015)

Electron Cyclotron Emission (ECE) and Correlation ECE diagnostics on TCV

Matteo **Fontana**[1,a], Laurie **Porte**[1], and Philippe **Marmillod**[1]

[1] *Swiss Plasma Center - EPFL - Station 13, Lausanne, CH*

Abstract. The Electron Cyclotron Emission (ECE) and correlation ECE diagnostics in TCV have been upgraded during its 2014 shutdown; this paper will provide a brief overview on their updated architecture and capabilities. The ECE system is equipped with two radiometers, each with 24 channels (750 MHz bandwidth) looking at the plasma both from the low and high field side covering the whole vessel. Of particular interest is the possibility of acquiring signals through a vertical line of sight and a dual-axis steerable antenna to study the electron distribution function (EDF) in plasmas with Electron Cyclotron Current Drive (ECCD) or EC Resonant Heating (ECRH). The correlation ECE system has been equipped with a new independent front end connected to the steerable antenna. The main characteristics of the radiometer are the six YIG independently tunable (between 6-18 GHz) 170 MHz bandwidth channels that can be moved on the whole LFS of the vessel. Using correlation analysis techniques it is possible to study very small temperature fluctuations. A brief presentation of some applications where these capabilities have been exploited in past TCV experiments for the study of micro instabilities characteristics is also included.

1 Introduction

Tokamak à Configuration Variable (TCV) is a medium sized machine for the study of the physics of magnetically confined plasmas for nuclear fusion applications. Its most relevant characteristics are the ability to work with a large range of different plasma shapes, its versatile heating system and the large variety of diagnostic systems available for the study of the plasma. In the past year the Electron Cyclotron Emission (ECE) diagnostic systems have been significantly upgraded. This kind of diagnostic allows the measurement of local electron temperature with high spatial and temporal resolution.

2 ECE

TCV has been equipped for the first time with ECE diagnostics in 2002 (HFS) and 2003 (LFS). In the following section we will describe the current status of the ECE systems on TCV.

2.1 Lines of sight

The ECE system of TCV is composed of two separate radiometers usually connected to a set of lines of sight looking from the HFS and the LFS, perpendicular to the magnetic field. Each radiometer can be connected to one of two antennas, one on the mid plane of the tokamak ($z = 0$) and the other at $z = 21$ cm. All of these transmit the signal up to the radiometers through an overmoded 1" circular waveguide ending in a rectangular section. The $z = 0$ LFS line of sight uses as a focusing element a polytetrafluoroethylene (PTFE) lens placed outside of the vessel port, while the one at $z = 21$ cm uses an in-vessel system composed by an elliptical mirror. Vacuum is assured by a quartz window. The HFS lines of sight are similar to the second LFS one with the difference that the waveguides run behind the tiles of the central column and bring the signal towards the upper part of the vessel where the vacuum break is provided by a 200 μm Mylar foil. The antenna pattern of these lines allows the study of modes with $k_\theta < 0.3\ cm^{-1}$, except for the LFS $z = 0$ one which is limited to $k_\theta < 0.7\ cm^{-1}$ at the vessel center.

2.1.1 Alternative lines of sight

In addition to these four lines of sight, TCV is equipped with a steerable receiver connected to the LFS and CECE radiometers, and a vertical receiver, whose signal can be transmitted to both HFS and LFS radiometers. Both lines of sight can be used to obtain information on the electron energy distribution function. The vertical line of sight consists of a focusing ellipsoidal mirror and a telescope system with a view from the top of the vessel, perpendicular to the magnetic field. A beam dump (reflectivity $< 0.1\%$) is placed at the bottom of the vacuum chamber. This line of sight can be considered to be looking along a surface of constant magnetic field. Neglecting the non-ideal antenna pattern and Doppler shifted emissions, different frequency components of the signal can be related to electrons with different kinetic energies (relativistic broadening) [1]. This antenna is connected with a mechanical switch both to the HFS and LFS radiometers allowing the use of both sets of channels [2].

[a] e-mail: matteo.fontana@epfl.ch

Another possible line of sight is a steerable antenna positioned at z = 45 cm (upper lateral). A detailed description of this antenna was provided by Goodman [3]. The movable mirrors allow one to rotate the line of sight with respect to a major radius (θ_L) and around it (ϕ_L), describing a cone trajectory. θ_L can be swept during a shot from 7 to 55° in 500 ms. This antenna will be shared among different diagnostics: doppler backscattering reflectometry, profile reflectometry, CECE and LFS ECE. The greatest advantages of this tool in ECE studies are the smaller spot size of the antenna (allowing the study of modes with $k_\theta < 1.12\ cm^{-1}$) and the possibility of using it to perform oblique measurements of the EC radiation. This is particularly interesting for plasma discharges where ECCD is present since it makes possible to study the difference between emission in co- and counter-current directions to get informations on the effects of current drive on the EDF.

2.2 Radiometer

The system has been designed to study second harmonic extraordinary mode (X2) electron cyclotron radition. After the circular to rectangular transition, the collected radiation passes through a couple of notch filters to block out the power from the X2 and X3 ECRH systems and then enters the two radiometers (whose scheme is presented in figure **??**). The incoming radiofrequency (RF) power is divided in a directional coupler and directed to two parallel heterodyne mixing stages, each connected to a different frequency local oscillator (LO): 63.9 and 82.5 GHz for the LFS radiometer, 76.5 and 94.5 GHz for the HFS one. Image-reject filters are placed before the mixing stages. In both radiometers the high frequency branch of the front end includes an isolator which prevents higher frequency LO power from the mixing stage from returning back to the lower frequency branch. The downconverted signals are then transmitted into low noise amplifiers (LNA) with 40 dB gain over the 2-20 GHz interval. These LNAs show 1 dB gain compression at 20 dBm (P1dB) and 3rd harmonic distortion after 30 dBm (IP3). After amplification each signal passes through a twelve-way power divider, splitting the IF into 24 channels, each characterised by a 750 MHz bandwidth band-pass filter. Each mixing stage is thus connected to 12 channels spaced by 1.5 GHz in the 2-20 GHz interval. Ultimately the RF bandwidths of the HFS and LFS radiometers are respectively 78-114 GHz and 66-100 GHz allowing us to make measurements over almost all the radial profile of TCV as it is shown in figure 2 (a) and (b).

The HFS radiometer has the additional possibility of choosing which RF signal to send into which set of channels (shifted 750 MHz one with respect to the other) through a couple of electronic switches, further increasing the flexibility in the choice of the resonant layers that can be analysed.

After being filtered the signals are rectified using a set of unbiased Schottky diodes (working in the 2-18 or 2-26 GHz interval), acquired at 200 kHz, low pass filtered (100 kHz) and amplified in a remotely controlled variable gain amplifier. The response curve of the diodes is non-linear up to -17 dBm.

2.3 Capabilities

The ECE system in TCV has been used in a wide range of plasma scenarios. Together, the HFS and LFS channels span the whole plasma minor radius. The 750 MHz bandwidth of an output channel corresponds to plasma volumes with different widths as the magnetic field varies over the radial profile: in the core, one channel corresponds approximately to a 1 cm wide region while in the HFS or in the LFS it would respectively be smaller and larger. The poloidal spot size varies for the different lines of sight limiting the maximum values of k_θ that can be resolved as mentioned before.

The choice of having two radiometers looking at the vessel from two different sides was made to allow a greater flexibility in the possible applications of the system. For example the LFS system makes it possible to study the bulk plasma temperature even in conditions where ECRH or ECCD generate a consistent fraction of suprathermal electrons: ECE radiation coming from this population, having a lower frequency than that of the correspondent emission layer due to relativistic effects, would be reabsorbed before reaching the LFS receivers. Instead, the HFS system can observe radiation emitted by these fast particles, giving the possibility to compare the results between LFS and HFS data coming from similar volumes in the plasma.[4]

3 Correlation ECE

The correlation ECE system in TCV was installed in 2007 but only with the last upgrade did it acquire an independent front end from the LFS ECE. Its working principle is to apply spectral decorrelation to ECE signals coming from plasma volumes very close to each other but not overlapped: if the distance between the volumes is chosen correctly (typically the turbulence correlation length), small scale modes will still be correlated while the thermal noise of the two channels will not. It is hence possible to lower the minimum fluctuation detectable by the radiometer ($\delta T_e/T_e$).[5]

3.1 Line of sight and radiometer

The newly upgraded CECE system in TCV uses the steerable receiver antenna. The RF stage of the CECE radiometer is similar to those, already discussed, of the standard ECE: two notch filters stop the ECRH power from reaching the mixing stage, a directional splitter divides the power between two mixers connected with LOs at 81.9 and 61.1 GHz, the lower side band is rejected and the down converted signals are amplified in LNAs with 30 dB gain over the 2-20 GHz interval and P1dB=10 dBm. Only one of the signals coming from the two mixers is transmitted to the IF stage and that can be manually switched between shots. The IF stage consists of a six-way power splitter

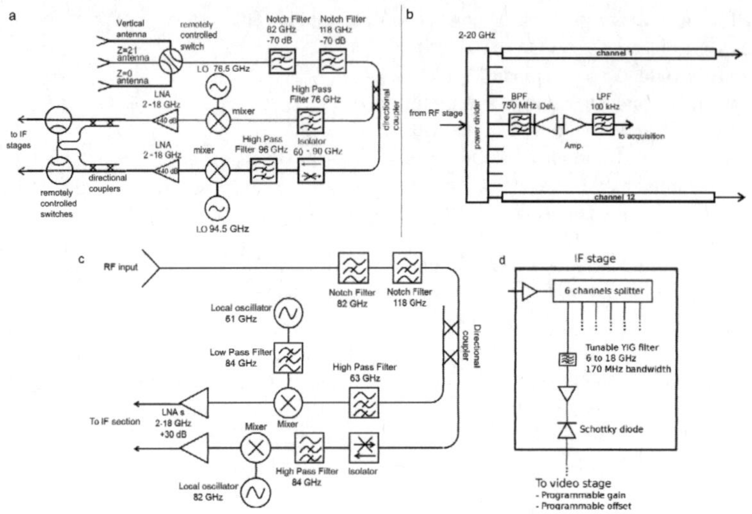

Figure 1. Schematics of the HFS (a RF stage, b IF stage) and CECE (c RF stage, d IF stage) radiometers. The LFS radiometer is identical to the HFS one except for the last section of the RF stage (the signal goes from the LNAs directly into the power dividers), the antenna connections and the working frequencies of LOs and other components.

leading to six YIG filters (170 MHz bandwidth), whose center frequency can be remotely adjusted between 6 and 18 GHz, each followed by another stage of amplification (20dB gain, 20 dBm P1dB, 27 dBm IP3) and an unbiased Schottky diode for rectification. The voltage signal is sampled at 1.67 MHz, low pass filtered at 450 kHz and then goes through video cards with programmable gains and offsets. A scheme of the CECE radiometer is presented in figure 1.

3.2 Capabilities

The newly upgraded CECE diagnostic is a highly versatile system: it can cover the 67-79 GHz or 88-100 GHz frequency intervals, depending on which mixer is being used, as shown in 2 (c), allowing to samle the whole LFS of the minor radius. Moreover the channels' frequencies, and hence the plasma volumes analysed, can be varied from one shot to another, further increasing the flexibility of the system. Having six simultaneous measurements allows to perform cross correlation analysis in up to three different plasma regions and estimate the correlation length of turbulent modes in a single shot. The radial resolution is mainly limited by relativistic effects, while in the poloidal direction by the antenna pattern as mentioned in 2.1.1.

3.3 Applications

The main field of application of the CECE is the study of microinstabilities and turbulence in tokamak plasmas, one of the causes of anomalous transport. The very narrow IF bandwidth of the YIG filters combined with correlation analysis allows the study of extremely low scale perturbations; the high filtering frequency in the video stage was chosen to properly take into account the effects of high

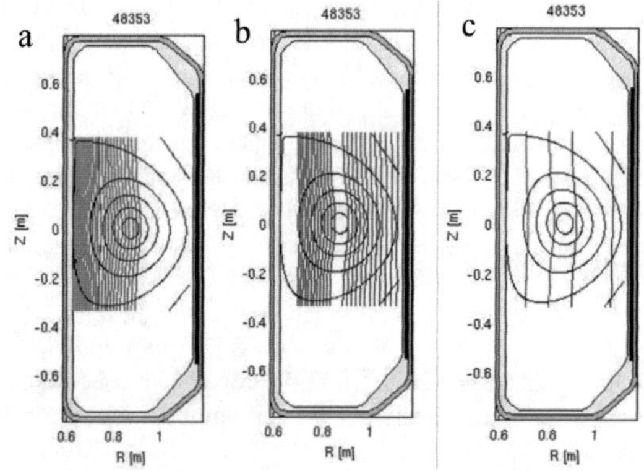

Figure 2. Representation of the resonance layers for the channels of (a) HFS ECE, (b) LFS ECE and (c) the two bands that can be accessed with the tunable YIG filters using the two different RF branches of the CECE.

frequency broadband turbulence. Finally the possibility to select the plasma volumes to analyse (by changing the center frequency of the YIG filters) over a wide radial range adds a further parameter that can be varied between similar shots to reconstruct profiles of turbulence characteristics.

In 2013 a series of experiments was performed with the goal of studying the effect of plasma shaping, in particular of the triangularity δ, on turbulence characteristics in discharges with constant density and collisionality. It was found that negative δ leads to a narrower turbulence frequency range and larger correlation length and that these effects tend to be more evident moving outward from the plasma core, since that is the region where the δ differ-

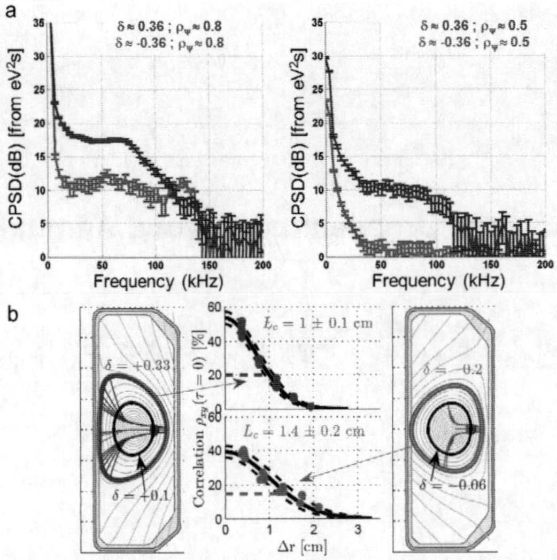

Figure 3. (a) Cross correlation spectra for two CECE channels at different radial positions (ρ) in shots with positive and negative δ (b) Correlation length values for shots with positive and negative δ.

ence is bigger [6], as shown in figure 3 . Given the upgrade of the CECE system, it is planned to expand the study of turbulence characteristics in a variety of operating conditions, including variations of density, shape, heating power, T_e/T_i (exploiting the recently installed 1 MW Neutral Beam Heating system), and regimes characterized by the presence of internal transport barriers. The results of these experiments will be compared with gyrokinetic simulations through the use of a novel CECE synthetic diagnostic.

Together with other diagnostics CECE has been used to characterize Geodesic Acoustic Modes (GAM) in TCV. These modes have been observed for $\rho_\psi > 0.7$ and its frequency matches the observation of other diagnostics such as magnetic coils or Tangential Phase Contrast Imaging (TPCI) [6], performed in different poloidal and toroidal positions

4 Conclusions

The ECE diagnostic is of fundamental importance to the study of plasma physics in magnetic confinement devices allowing the possibility of obtaining information on local electron temperature with very high spatial and temporal resolution. TCV is equipped with a very flexible system that is able to cover the whole plasma cross section looking from the HFS and from the LFS. An important asset of this system is the presence of two non conventional lines of sight, one looking at the plasma along a vertical direction, while the other is connected to an antenna steerable around two axes, one of which can be scanned during shots. TCV is also equipped with a CECE system that can be used to study microfluctuations and turbulence characteristics at very low spatial scale by means of spectral decorrelation analysis. Future work will take advantage of the improved capabilities of the CECE diagnostic to study electron temperature fluctuations in a broad range of operating conditions.

Acknowledgements

This project has received funding from the HGF Virtual Institute on Plasma Dynamical Processes and Turbulence Studies using Advanced Microwave Diagnostics. This work was also supported in part by the Swiss National Science Foundation.

References

[1] T.C. Luce, P.C. Efthimion, N.J. Fisch, Review of Scientific Instruments **59**, 1593 (1988)

[2] W.W. Eshetua, L. Porte, A. Fasoli, O. Sauter, S. Coda, T.P. Goodman, The TCV Team, EPJ Web of Conferences **32**, 03011 (2012)

[3] T.P. Goodman, V.S. Udintsev, I. Klimanov, A. Mueck, O. Sauter, C. Schlatter, FST **53**, 196 (2008)

[4] I. Klimanov, L. Porte, S. Alberti, P. Blanchard, A. Fasoli, T.P. Goodman, Review of Scientific Instruments **76**, 093504 (2005)

[5] C. Watts, FST **52**, 176 (2007)

[6] L. Porte et al., in *Proceedings of 25th IAEA Fusion Energy Conference, St Petersburg, Russia* (2014)

Experimental Measurement of ECH Deposition Broadening: Beyond Anomalous Transport

M. W. **Brookman**[1,a], M. E. **Austin**[1], K. W. **Gentle**[1], C. C. **Petty**[2], D. E. **Ernst**[3], Y. **Peysson**[4], J. **Decker**[4], and K. **Barada**[5]

[1] *Institute For Fusion Studies, University of Texas at Austin*
[2] *General Atomics, San Diego, CA*
[3] *PSFC - MIT*
[4] *CEA - Cadarache*
[5] *UCLA*

Abstract. This work provides a first experimental measurement of broadened ECH deposition on the DIII-D tokamak. As seen in theory[1] and simulation[2], refraction by edge density fluctuations shifts the path of RF waves, altering ECH and ECCD deposition. This paper reports on an initial experimental confirmation of broadened ECH deposition on DIII-D tokamak. T_e measurements from a 48 channel 2nd Harmonic ECE Radiometer digitized at 500 kHz are used with a set of broadened trial ECH deposition functions to calculate time-dependent, modulation-induced heat fluxes. The fitting of convective and diffusive transport to these fluxes allows different ECH deposition profiles to be compared. The best-fit ECH deposition produces reasonable transport coefficients which compare favorably with simulation. This method is applied to a set of L- and H- mode DIII-D discharges. Accounting for diffusive, convective, and coupled transport, the ECH deposition profile is found to be 2 to 3 times wider than predicted by TORAY-GA ray tracing.

1 Introduction

Injection of RF power at the electron cyclotron frequency is used for localized heating (ECH) and current drive(ECCD) in tokamaks. EC deposition has been treated by 1D ray tracing and more complex full-wave calculations[3] and benchmarked against experimental deposition measurements[4]. Still, unresolved questions with significance for ITER remain. Anomalous transport has been credited for the discrepancy between ray tracing predictions of ECCD profiles and experimentally measured profiles, which are spatially broader [5] [6]. Work at TCV on ECCD [7] casts doubt on anomalous transport as a primary driver of broadening. Certain experimental signatures are expected from the fast electrons generated by ECCD. However, synthetic HXR on TCV is inconsistent with a significant diffusion of EC driven current.

Instead, turbulent edge density fluctuations, present in all tokamak plasmas, can defract RF waves traveling through them. Edge $\tilde{n}_e$ effects on RF propagation have been seen in theory[1] and computation [2]. These fluctuations are located in the edge ($\rho > .9$) and are driven by drift wave turbulence. They have a characteristic size on the order of $\sim 5mm$ and amplitudes on the order of $\tilde{n}_e/n_e \sim .1$. Fluctuations interact with a range of frequencies, lower hybrid[8] with f∼5 GHz, as well as EC waves with f∼ 100 GHz. This edge interaction changes the in-

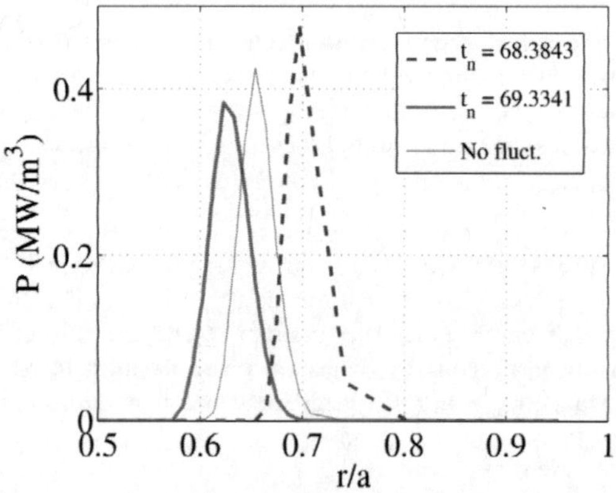

Figure 1. Results from Decker on TCV showing the ECH source function's range of variation over time due to edge fluctuations. Sufficiently rapid shifts compared to observation frequencies cause an effectively broader source function[7]/

tersection of the RF beam with the EC resonance on an edge turbulence timescale, $\tau < 1ms$. Fig. 1 shows an example of the defraction effect on the $P_{ECH}(r)$ deposition profile at different points in the same discharge. Observed by diagnostics, which have some finite response time, this appears as an effectively broadened profile. DIII-D experi-

[a]e-mail: brookmanmw@fusion.gat.com

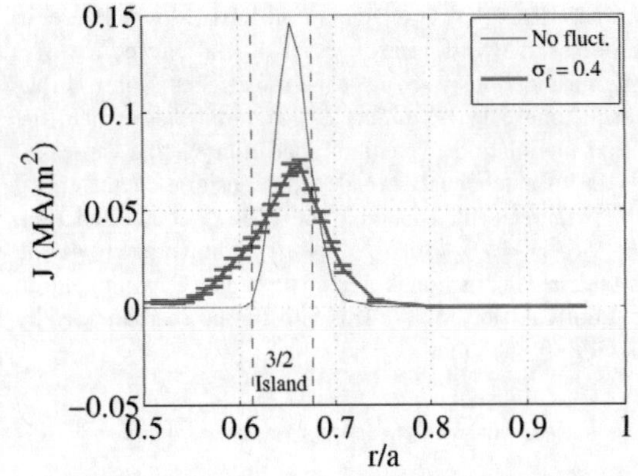

Figure 2. ECCD profiles predicted for ITER from the code C3PO/LUKE in the presence of modest edge fluctuation levels $\sigma_f \propto \tilde{n}_e/n_e$. Reprinted with kind permission of The European Physical Journal [7]

ments have also seen indications of broadening on the 110 GHz 2nd harmonic X-mode gyrotron system. Modulated ECH transport experiments by Gentle[10] found a hysteresis effect in critical gradient behavior, but this is minimal if $P_{ECH}(r)$ deposition is taken as ~ ×2 wider than predicted by TORAY-GA.

ECCD and ECH experimental profile measurements are made through modulation, perturbing a discharge with a known input, and analyzing its effect on plasma parameters[4]. ECCD measurements can be made by Motional Stark Effect(MSE)[6]. MSE measures the change in pitch angle of magnetic field lines when ECCD is switched from co- to counter-current drive. This change in pitch can be analyzed to produce a driven current profile. The resistive timescales which govern current profiles limit ECCD measurement to slower modulations, no faster than 20 Hz. This gives transport processes more time to obscure the instantaneous deposition profile, and because ECCD acts on a specific set of particles in phase space, this transport can be in both space and momentum [9] .

Measurement of ECH deposition through ECE is significantly simpler. Heat transport in tokamak plasmas can be treated in a radial 1D model[10]. Higher time and spatial resolution temperature measurements allow for faster modulation to be used, up to 1 kHz. Given that edge fluctuation broadening alters ECCD by means of altering ECH deposition location, ECE heat transport measurements offer a means of understanding this effect without the complexities of ECCD measurement.

It is vital to understand the impact of edge broadening on ITER. Altered ECCD deposition could hinder NTM suppression efforts, as the machine will need nearly its full power to suppress the predicted 3/2 mode[11]. For a modest level of edge fluctuations, the LUKE/C3PO code (Fig. 2) predicts this effect will cause a significant amount of ECCD to fall outside the 3/2 island. Current driven outside the island does not contribute to mode suppression.

1.1 Overview

In a series of discharges in the DIII-D tokamak an analysis of electron temperature data from ECE provides information about ECH deposition. For a best fit with linear coefficients, a significant level of deposition broadening is needed. This work explores a flux integral method of treating transport, performed in the Fourier basis of a modulated ECH power. A best fit $P_{ECH}(r)$ is found based on the relation between transport drive terms and heat flux. The flux is calculated (Sec. 2), and the method of transport fitting explored (Sec. 3.1). It is applied to set of DIII-D discharges with varying levels of edge turbulence (Sec.4.1). A 1.7 to 2.8 times broader deposition produces best fit and most reasonable transport coefficients.

2 Resolving Deposition, Transport through Flux

2.1 1D Heat Equation

To understand deposition, it is necessary treat the time-dependent heat flux. Perturbed by heating, transport moves ECH power promptly outside its initial deposition region. The 1D heat equation [13] provides the basis for calculating heat flux. It relates local stored energy of the plasma to heat and particle fluxes, sources, and sinks. Here P_{ECH} is the ECH source function and P_{OTHER} includes contributions from e-i exchange, radiative, ohmic, and neutral beam power.

$$\frac{3}{2}\frac{\partial}{\partial t}(nT_e) + \nabla \cdot (Q_e + \frac{5}{2}\Gamma T_e) = P_{ECH} + P_{OTHER} \quad (1)$$

The use of 50% duty cycle square wave modulated ECH power created a periodic change in plasma temperature. Linearizing the heat equation in this perturbation for fast modulation, f_{mod} > 20 Hz produces a simplified form. P_{OTHER} is not directly modulated. Neutral beam power is constant and other indirectly modulated terms are small in amplitude according to ONETWO simulation of the discharges in this work. Based on Thomson scattering and reflectometry density measurements, for fast modulation $\tilde{T}_e/T_e >> \tilde{n}_e/n_e \approx 0$. Assuming the particle flux is negligible, a heat flux can be calculated from an experimental temperature and an ECH deposition profile.

$$\tilde{Q}_e(r,t) = \frac{1}{r}\int_{r'=0}^{r}(-\frac{3}{2}n_e(r')\frac{\partial \tilde{T}_e}{\partial t}(r',t) + \tilde{P}_{ECH}(r',t))r'dr \quad (2)$$

2.2 Fourier Analysis of ECE Temperature

In the absence of transport (i.e $\tilde{Q}_e = 0$), the local power deposition will be equal to the change in stored energy of the plasma, a time derivative of T_e. Rather than treat this derivative directly in experimental data, it is preferable to use Fourier analysis to produce a series of harmonic equations which reflect the time dependence. Expressing a Fourier transformed quantity with the 'hat' notation,

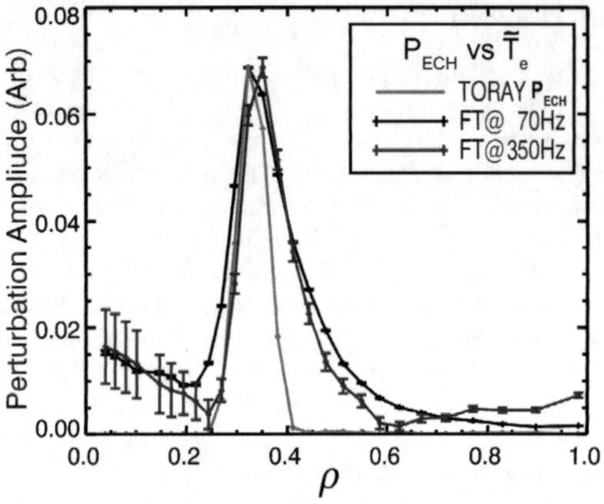

Figure 3. Normalized $\hat{T}_e$ perturbation profile produced with error bars by Fourier Analysis. 1st (black) and 5th (blue) harmonics. These are compared with the $P_{ECH}(r)$ source function predicted by the TORAY-GA ray tracing code [3].

$$\frac{3}{2}n_e\frac{\partial \tilde{T}_e}{\partial t}(r,t) \approx \tilde{P}_{ECH}(r,t) \xrightarrow{DFT} \frac{3i\omega}{2}n_e\hat{T}_e \approx \hat{P}_{ECH}(r,f) \tag{3}$$

T_e data comes from a 48 channel 2nd Harmonic X-mode ECE radiometer digitized at 500 kHz. A least-squares fit Fourier series converts 48 $T_e(\rho,t)$ signals over a 500 ms time period into a complex harmonic series in $n \times f_{mod}$. Odd harmonics dominate due to the square-wave input power. Faster timescale behavior is captured by higher frequencies, and thus the time dependence of the heat flux can be discerned. Error is propagated through the fit uncertainty of the Fourier harmonics, and reflectometry-derived density, and the ECH source function terms.

Fig. 3 shows the 1st and 5th harmonic of the temperature perturbation as compared to a TORAY source function. The TORAY profile is substantially narrower, due to a combination of deposition broadening and transport. These will be distinguished through treatment of the time-dependent flux in the Fourier basis of the perturbation[14].

Diffusive heat flux is driven by the temperature gradient, $\nabla\hat{T}_e$ using a high tension B-Spline, this term can be calculated from the perturbation profile. Fig. 4 shows ECE T_e data from the 5th harmonic. A Monte Carlo integral can directly be taken on each part of the complex $\hat{T}_e$ and $\hat{P}_{ECH}$ expressed in a real and complex 'x+iy' formulation. An amplitude and phase '$Ae^{i\phi}$' integral has a significantly more complex form.

2.3 ECH Source Functions

The heat flux depends on experimental parameters, but also requires some assumed form of the input power. Past ECH transport work has often used a rectangular radial deposition function for analytic simplicity[17], but ray-tracing TORAY-GA, as well as C3PO/LUKE code from TCV[7], predict a smoother deposition. In lieu of LUKE/C3PO simulations for DIII-D, which are in progress,this work relies a series of trial blurred $P_{ECH}(r)$ source functions. As the edge effect occurs on a turbulence timescale 1-2 orders of magnitude faster than the ECH modulation. Thus the broadening can be expressed as an instantaneously broadened deposition characterized the parameter 'b', applied to the TORAY source function, $P_{ECH}(\rho,t) = P_{ECH}(\rho,b) \times P_{ECH}(t)$. Shown in Fig 4., the broadened deposition is created through a Gaussian blur in cylindrical coordinates. This blur can be characterized by a width b, such that

$$P_{ECH}(\rho,b) \propto \rho^{-1}(\rho P_{TORAY}(\rho) * exp(\frac{-\rho^2}{2b^2})) \tag{4}$$

ECH input power does not form a perfect square wave. Gyrotrons have a finite switching time, and phases which vary by a few degrees, and thus the ECH does not posses a strictly $sin(n2\pi f)/n$ character and the form of $P_{ECH}(t)$ must be considered. By performing the same Fourier series fit on the measured forward power from the gyrotrons, this flux integral fitting accounts for space and time variation of ECH in this flux calculation. This broadening and Fourier analysis are applied independently to each of the six gyrotrons to produce $\hat{P}_{ECH}(\rho,f)$ for calculating heat flux.

2.4 Heat Fluxes Indicative of Broadening

High-power ECH heating generates both an inward and outward traveling heat flux[17]. This constrains the phase and amplitude of $\hat{Q}_e$, and even before fitting transport the flux can help constrain the width of P_{-ECH}. Amplitude is small near the deposition center. Moving out from this region, a phase lag in $\hat{Q}_e$ reflects the time it takes for heat to propagate. Fig 6 shows a comparison of the amplitude and phase of the fundamental harmonic of $\hat{Q}(\rho,f)$ for various broadened input source functions- a TORAY source functions, a 2.3 times wider FWHM case and over-broadened 2.8 times wider case. In the next section, a 2.3 factor will be found to produce the best fit, and the corresponding flux shows a monotonic phase lag around the deposition. Higher frequencies reflect faster behavior, so higher harmonics should experience less transport and have a larger phase lag. Fig 7 shows a H-mode case with the correct broadening level. This H-mode has a significant ELMing character which perturbs the edge heat flux, but this perturbation does not reach into the core.

3 Transport Fitting and Validation

3.1 Selecting Transport Coefficients

Resolving ECH deposition requires treatment of transport. Convective and diffusive parts are needed to explain the heat flux induced by ECH[19]. $Q_e(\rho,t) = Dn_e\nabla T + Vn_eT$. Thus a modulated heat flux, $\tilde{Q}_e$, must be expressed as a linearizion in each parameter, including the transport coefficients.

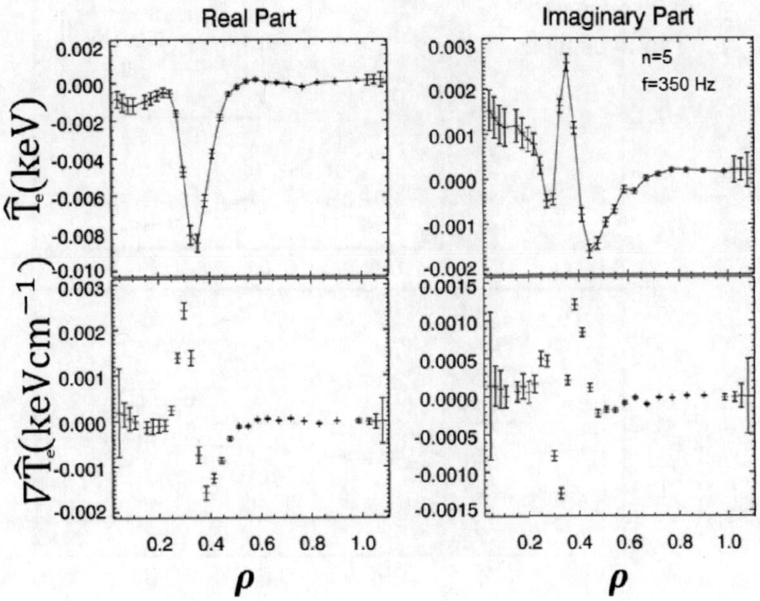

Figure 4. Real, imaginary parts of the perturbed temperature $\hat{T}_e$ as well as its gradient. The 5th harmonic of the perturbation is shown, higher harmonics have larger uncertainties. Up to the 17th harmonic, 1050 Hz, can be resolved in this discharge.

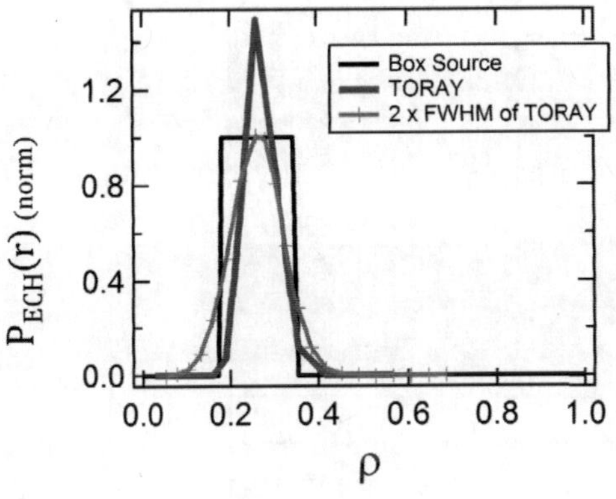

Figure 5. Comparison of possible source functions with the same total power. C3PO/LUKE calculations are not yet available for DIII-D, so broadened TORAY source functions are used.

$$\tilde{Q}_e = n_e(D\nabla\tilde{T}_e + \tilde{D}\nabla T_e + V\tilde{T}_e + \tilde{V}T_e) + \tilde{n}_e(D\nabla T + VT) \quad (5)$$

The time dependence of the modulated transport coefficients is indirect, entering through other plasma parameters which are periodically perturbed. Expressing this as an expansion for $\tilde{D}$:

$$\tilde{D} = \nabla\tilde{T}_e\frac{\partial D}{\partial\nabla T_e} + \tilde{T}_e\frac{\partial D}{\partial T_e} + \tilde{n}_e\frac{\partial D}{\partial n_e} + \nabla\tilde{n}_e\frac{\partial D}{\partial\nabla n_e} \quad (6)$$

The terms which make up the modulated flux can be inserted and grouped into three distinct parts in the form $C2\nabla\tilde{T}_e + C1\tilde{T}_e + C0$. [13]

$$\frac{\tilde{Q}_e}{n_e} = -\nabla\tilde{T}_e(D+\nabla T_e\frac{\partial D}{\partial\nabla T_e})+\tilde{T}_e(V+T_e\frac{\partial V}{\partial T_e}-\nabla T_e\frac{\partial D}{\partial T_e})$$
$$+\nabla\tilde{n}_e(T_e\frac{\partial V}{\nabla n_e}-\nabla T_e\frac{\partial D}{\nabla n_e})+\tilde{n}_e(T_e(\frac{V}{n_e}+\frac{\partial V}{\partial n_e})-\nabla T_e(\frac{D}{n_e}+\frac{\partial D}{\partial n_e})) \quad (7)$$

Consider $D_M = D + \nabla T_e\frac{\partial D}{\partial\nabla T_e}$ as a modulated diffusion driven by $\nabla\tilde{T}_e$. Likewise, $V_M = V + T_e\frac{\partial V}{\partial T_e} - \nabla T_e\frac{\partial D}{\partial T_e}$ is a modulated convection. The remaining term - driven by the modulated density and its gradient - will be grouped together as coupled transport, $\tilde{\xi}$. These simplifications produce a fitting equation of the form:

$$\hat{Q}_e(r,f) = n_e D_M\nabla\hat{T}_e + V_M n_e\hat{T}_e + \hat{\xi}$$
$$= \frac{1}{r}\int_{r'=0}^{r} 3i\pi f n_e(r')T_e(r',f) + P_{ECH}(r')\hat{P}_{ECH}(f)r'dr \quad (8)$$

3.2 Treating Modulated Transport

Constant diffusion and convection coefficients are insufficient to describe the full behavior of the time-dependent flux. Additional effects, collectively referred to as coupled transport, provide an additional contribution to the flux.

 DOI: 10.1051/epjconf/201714703001

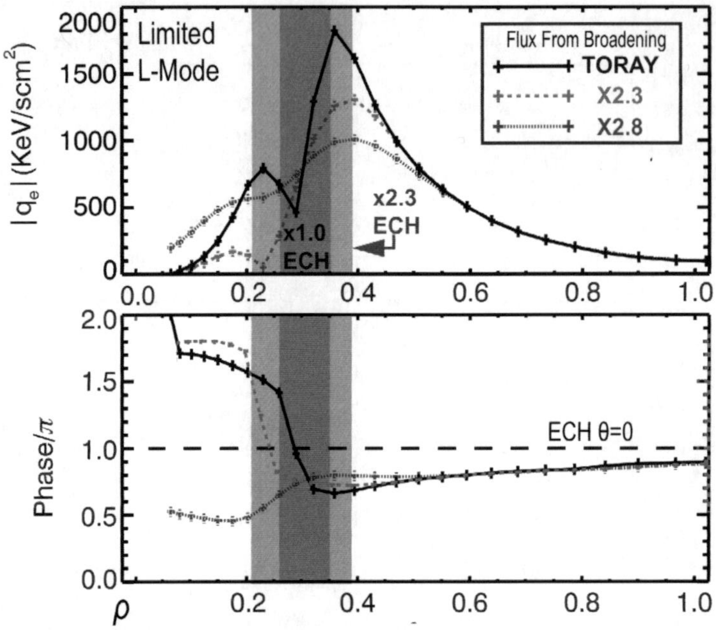

Figure 6. First harmonics of differently broadened heat fluxes from a limited L-Mode. A TORAY width source function overshoots in its phase change, as an artificially high flux from the narrow deposition region is needed to match the measured temperature perturbation. An overly broadened source, x2.8 wider, shows a very low phase in the core as an aphysical flux into the region is needed to correct the assumed P_{ECH}. A x2.2 wider source function, later found to be the best-fit, has a monotonic phase change across the deposition zone.

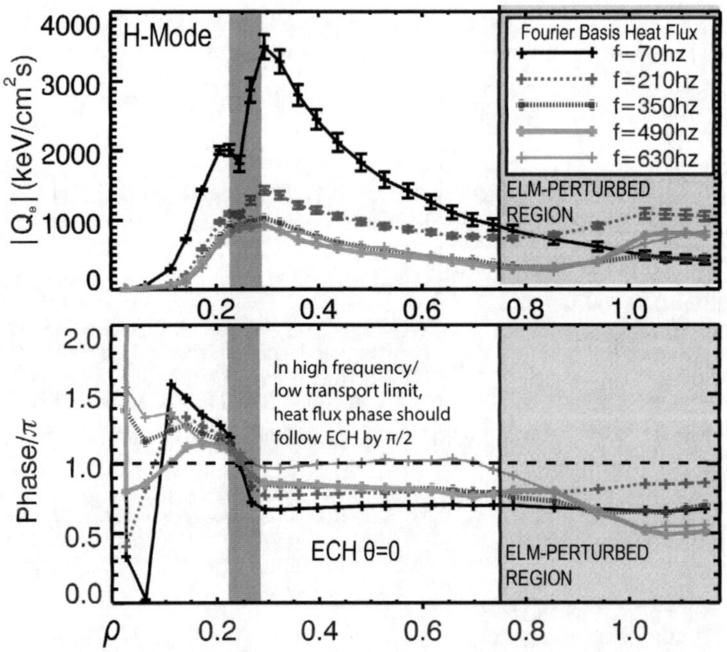

Figure 7. Heat flux calculated for a set of harmonics in an H-mode discharge with the best-fit source function. Inwards and outwards traveling heat pulses are clear from the phase at center of the deposition. The order of the phases inverts as lower frequency reflect slower changes.

Since its drivers are not known a-priori, it must be treated as a fit parameter. $\hat{\xi}$ is a complex term with an unknown frequency dependence. Fitting to a variety of discharges has found the ∇T_e driven portion of $\hat{\xi}$ dominates. This makes $\hat{\xi}$ reminiscent of an anomalous diffusion[16].

$$\hat{\xi} \approx -\nabla T_e(\nabla \tilde{n}_e \frac{\partial D}{\nabla n_e} + \tilde{n}_e(\frac{D}{n_e} + \frac{\partial D}{\partial n_e}))$$

Because fitting will be performed across harmonics, some form of time dependence must be applied, as $\nabla \tilde{n}_e$ and $\tilde{n}_e$ are not known. This is taken to be an exponential in time over a single modulation cycle -

$$\tilde{\xi}(t) \propto \xi_0 e^{-c(t-t_0)}, \{t = (t_0, t_0 + 1/f)\} \quad (9)$$

This has a simple Fourier transform, which is the same up to a complex factor as that of the double-sided exponential decay. Thus we proceed with the form

$$\hat{\xi} = \frac{C_0 c e^{i\theta_0}}{c^2 + 4\pi^2 f^2} \nabla T_e \quad (10)$$

The 3 fit parameters for this form are then an amplitude C_0, a phase θ_0, and an exponential decay time c. There are three justifications for this form of diffusive-like coupled transport. Firstly, an exponential modulated diffusion appears frequently in the literature to fit heat pulse transport in experiments[16]. Time variation of $\hat{\xi}$ can be expressed in an exponential basis as well as a sine wave basis

$$\hat{\xi} = \sum_n a_n exp(-nt) \quad (11)$$

Adding a full series of fit exponentials with variable phases would introduce a massive number of non-orthogonal fit parameters. Instead, this work fits a single decaying exponential form. A χ^2 minimization will find the single frequency response which best accounts for time variation not driven by D_M and V_M. Furthermore, this form seems to produce the best fit results. The imposed form of frequency fit produces a smaller χ^2 result than a triangle or sawtooth wave time dependence, $\hat{\xi}(f) \propto 1/f^2$ or $\hat{\xi}(f) \propto 1/f$.

3.3 Performing the Fit

Equ. 5 has five unknown parameters - D_M, V_M, and the 3 parameters of $\hat{\xi}$ are fit to the sets of heat fluxes calculated for each of the trial P_{ECH} deposition profiles. This avoids the challenge of conserving the total input power in a fit parameter form of P_{ECH} which includes radial variation. Fits are performed using the uncertainties propagated in each parameter using orthogonal distance regression, ODR[18]. A traditional least-squares fit uses variation of an independent variable, here $\hat{Q}_e/n_e$, to minimize the quadrature difference between the calculated and fit values. The independent variables, ∇T_e, $\nabla \hat{T}_e$, and $\hat{T}_e$ would need to be taken as exact. When fitting the heat equation both the independent and dependent variables have substantial known errors. Q_e and its uncertainty are Monte Carlo integral quantities, not strictly local. Errors in $\nabla \hat{T}_e$ are larger than those in $\hat{T}_e$, but are of a similar magnitude. To treat these errors, ODR varies dependent variables as well, minimizing the uncertainty-weighted orthogonal separation between the data points.

The fit must also account for radial variation. Transport coefficients can change by an order of magnitude or more across the Tokamak. In the next section, results are presented from both locally constant fits made over a limited range of ρ and from polynomial fit coefficients. Benchmarking shows that these two methods match within their uncertainties.

4 Experimental Characterization of Broadening

4.1 χ^2 Minimization & Transport Coefficients

This method has been applied to a set of discharges on the DIII-D Tokamak. The first of these was a wall-limited L-mode discharge where substantial broadening was expected. Fig. 8 shows a series of 5th order polynomial fits of the radial variation of diffusion coefficients, and the resulting χ^2 of the fit. The minimization of χ^2 occurs for a source function with a FWHM x2.3 broader than predicted by TORAY-GA. The best fit source function produces a reasonable diffusion which is similar to the predicted diffusivity from the ONETWO transport code.

4.2 Turbulence Comparison Through Matched H- and L- Mode Discharges

Modulated ECH was injected into a series of DIII-D discharges in shape and density matched H- and L-mode discharges. Deposition was targeted at two locations, ρ = 0.25, and ρ = 0.6. These targets have a significantly different path length through the plasma, but both generate inward and outward heat pulse propagation.

Two power levels, 3 MW and 1.2 MW, were applied with $f_{mod} = 70Hz$. Density and deposition location are matched in the shapes, but improved H-mode confinement leads to a higher temperature. This broadens the ECH resonance, but this effect is already treated by the TORAY-GA ray tracing code. Edge turbulence drops with the formation of the H-mode pedestal[20], allowing for a comparison of broadening in the same shape at two turbulence levels. Deposition and plasma profiles, as well as the Doppler backscattering channel measuring edge fluctuations for two matched core heating discharges are shown in Fig. 10. Transport analysis on these discharges has been performed, with a best-broadening polynomial H-mode fit to D_M and V_M is shown in Fig. 9 for comparison with the earlier L-mode polynomial fit. D_M compares very well with the χ_e diffusivity from ONETWO in the region where V_M is small - where transport is dominated by diffusion and $\chi_e \approx D_M$. The ELM effects on the H-mode flux lead to unreasonable values of D_M and V_M in the edge.

To compare these discharges, constant coefficients were fit in the range ρ = (.3, .6) for a series of broadening factors. The H-mode discharge shown in Table 1 and

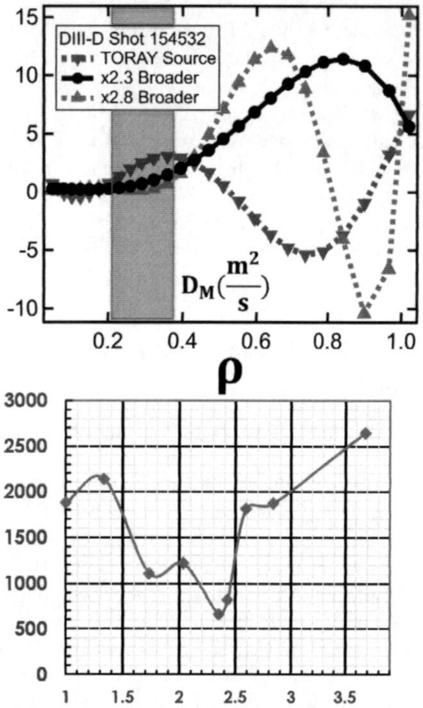

Figure 8. Physical values of D_M for a Limited L-mode coincide with minimization of the χ^2. A 5th order polynomial can be used to represent radial variation in the Diffusion Coefficient (above). This can be compared to the ONETWO power balance diffusivity. The best fit occurs for a ×2.3 broader FWHM for $P_{ECH}(r)$. This fits a consistently positive diffusion coefficient similar to the ONETWO value.

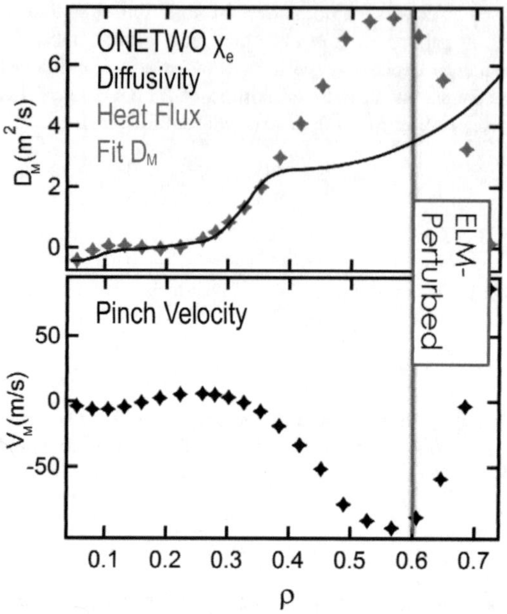

Figure 9. H-mode best fit transport coefficients are reasonable inside of $\rho = .6$. Outside this radius the effect of ELMs overwhelms the ECH heat flux, especially at high harmonics.

Broad. Factor	$D_M(cm^2/s)$	$c(1/s)$	χ^2
×1	3.2 ± .1	747	4067
×1.2	1.2 ± .1.4	2191	1223
×1.5	1.4 ± 1	877	916
×1.7	2.7 ± 1.3	306	**223**
×2.3	5.4 ± .8	7025	393

Table 2. Diverted H-mode discharge fits best from $\rho = (0.3, 0.6)$ with a modest ×1.7 broadening. Diffusion is comparable across cases to the ONETWO value for effective diffusivity $\chi_{eff} = 1.7 - 2.5$

the L-mode in Table 2 show a clear minimization of χ^2 and reasonable transport coefficient for broadened profiles, with L-mode requiring significantly more broadening than H-mode. Also shown in these tables are the characteristic times, c, which define the best fit exponential decay. The two best-fit cases produce very similar characteristic timescale for coupled transport at the best-fit broadening factor.

Broad. Factor	$D_M(cm^2/s)$	$c(1/s)$	χ^2
×1	1.9 ± .4	423	2223
×2.0	4.7 ± .5	386	520
×2.5	4.6 ± .4	321	588
×2.75	3.5 ± .2	293	**346**
×3.2	3.6 ± .2	-2844	555

Table 1. The diverted L-mode fit from $\rho = (0.3, 0.6)$ with a much larger broadening factor multiplied onto the very narrow TORAY source. The FWHM of P_{ECH} is still less than the limited L-mode above.

5 Conclusions and Future Work

Calculation and fitting of modulated heat flux is a promising means of resolving ECH deposition. Best-fit transport coefficients are concurrent with significant ECH deposition broadening in both H- and L- mode predicted by TORAY-GA. For discharges which have a higher level edge turbulence, a substantially broader ECH deposition is found. These results are summarized in Table 3.

In future work, a causal connection between fluctuation levels and broadening factors will be explored, either through adjustment of edge fluctuations or a rigorous elimination of anomalous transport effects by adjusting targeting and plasma parameters. Other confinement modes, such as QH-mode and a recently achieved DIII-D negative triangularity L-mode, offer different edge turbulence conditions to be explored. LUKE/C3PO Simulations are underway for DIII-D, to produce computationally broadened profiles for use with the heat flux method.

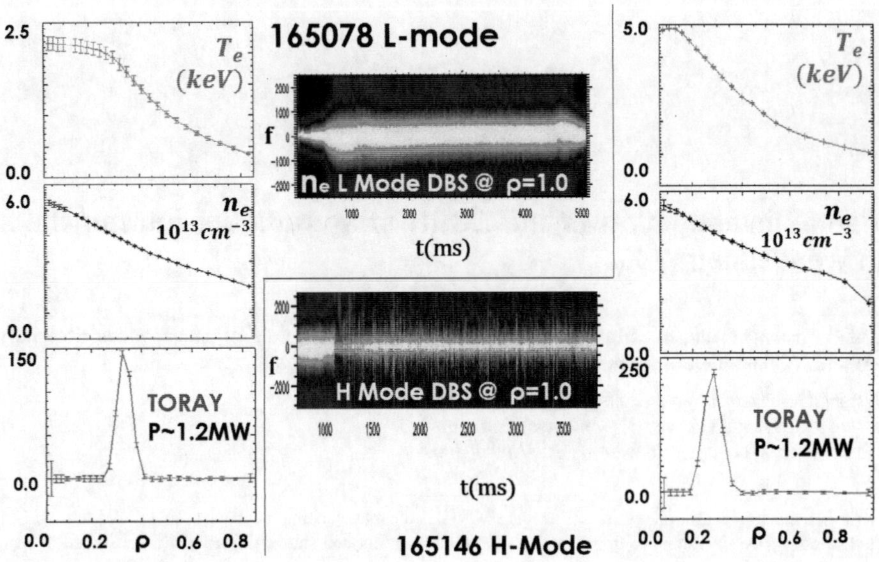

Figure 10. Profiles for the matched shape diverted H- and L- mode shots. The H-mode shows a clear reduction in edge density fluctuations as measured by DBS, which should lead to a drop in ECH broadening.

Discharge:	Limited L-mode	Diverted L-mode	Diverted H-mode
TORAY FWHM	0.072	0.055	0.067
Best Fit FWHM	0.17	0.15	0.11
Factor	×2.3	×2.7	×1.7

Table 3. FWHM of the best fitting broadened $P_{ECH}(r)$ from different shots. These discharge modes should have a steadily decreasing level of edge turbulence.

References

[1] Kyriakos H. & Ram A.K. Phys. Plasmas **17**, 022505 (2010)
[2] Peysson Y. et Al Plasma Phys. Control. Fusion **53**, 124028 (2011)
[3] Prater R. et Al, Nucl. Fusion **48**, 035006 (2008)
[4] Zerbini M. et Al, Plasma Phys. Control. Fusion **41**, 931 (1999)
[5] Kirov K.K. et Al, Plasma Phys. Ctrl. Fusion **44**, 2583 (2002)
[6] Petty C.C. et Al, Nucl. Fusion **42**, 1366 (2002)
[7] Decker J. et Al, EPJ Web of Conf. **32**, 01016 (2012)
[8] Peysson Y. & Decker J. F.S.T 65, 22 (2014)
[9] P. Nikkola et Al, Nucl. Fusion **43** 1343 (2003)
[10] Gentle K.W. et Al, Phys. Plasmas **13**, 012311 (2006)
[11] LaHaye R.J. et Al, ECA **25A**, 1357 (2001)
[12] Harvey R.W. et Al, Phys. Rev. Lett. **88**, 205001 (2002)
[13] Gentle K.W. et Al, Phys. Fluids **31**, 1105 (1988)
[14] Stockdale R.E. et Al, Bull. Am. Phys. Soc. **31**, 1535 (1986)
[15] Peysson Y. et Al Plasma Phys. Control. Fusion **54**, 045003 (2012)
[16] Fredrickson E. D., Phys. Rev. Lett. **65**, 2869
[17] DeBoo J.C. et Al, Phys. Plasmas **19**, 082518 (2012)
[18] Boggs P.T. et Al, Contemporary Mathematics **112**, 186 (1990)
[19] Mantica P. et Al, Phys. Rev. Lett. **95**, 185002 (2005)
[20] Rettig C. L. et Al, Nucl. Fusion **33** 643 (1993)

This material is based upon work supported by the U.S. Department of Energy, Office of Science, Office of Fusion Energy Sciences, using the DIII-D National Fusion Facility, a DOE Office of Science user facility, under Award DE-FC02-04ER54698 and DE-FG03-97ER54415. DIII-D data shown in this paper can be obtained in digital format by following the links at https://fusion.gat.com/global/D3D_DMP

Experimental investigation of the ECRH stray radiation during the start-up phase in Wendelstein 7-X

Dmitry **Moseev**[1,a], Heinrich **Laqua**[1], Stefan **Marsen**[1], Torsten **Stange**[1], Harald **Braune**[1], Volker **Erckmann**[1], Florian **Gellert**[1,2], Johann Wilhelm **Oosterbeek**[3], Uwe **Wenzel**[1], and the W7-X Team

[1] *Max-Planck-Institut für Plasmaphysik, Greifswald, Germany*
[2] *Ernst-Moritz-Arndt-Universität Greifswald, Greifswald, Germany*
[3] *Eindhoven University of Technology, Eindhoven, The Netherlands*

Abstract. Electron cyclotron resonance heating (ECRH) is the main heating mechanism in the Wendelstein 7-X stellarator (W7-X). W7-X is equipped with five absolutely calibrated sniffer probes that are installed in each of the five modules of the device. The sniffer probes monitor energy flux of unabsorbed ECRH radiation in the device and interlocks are fed with the sniffer probe signals. The stray radiation level in the device changes significantly during the start-up phase: plasma is a strong microwave absorber and during its formation the stray radiation level in sniffer probes reduces by more than 95%.
In this paper, we discuss the influence of neutral gas pressure and gyrotron power on plasma breakdown processes.

1 Introduction

Wendelstein 7-X (W7-X) is an optimized stellarator with $n = 5$ symmetry operating in Greifswald, Germany. Electron cyclotron resonance heating (ECRH) is the main heating mechanism in the device [1–3]. In its first experimental campaign, OP1.1, it was equipped with 4 MW of ECRH power as a sole source of heating. Plasma was started also by means of ECRH.
In the highly reflective vacuum vessel of W7-X, unabsorbed microwave radiation can reach high energy fluxes (up to 350 $kW/m^2/MW$ [4]) when there is no plasma. Therefore, studying the start-up processes is of high importance for the safety of the machine and its diagnostics. Both helium and hydrogen were used as fueling gasses for the experiments. The first plasma was produced in helium on 10.12.2015 and the fueling gas was changed to hydrogen on 03.02.2016. Plasma breakdown times for the two gasses are discussed in this contribution.
In Section 2 plasma breakdown is defined; the influence of neutral gas density and power on the delay time and the density evolution are shown in Section 3; Section 4 concludes the paper.

2 Definition of plasma start-up time

The plasma start-up time is usually defined as time from the beginning of heating, needed for theH_α emission to reach its maximum. It was shown [5] that the maximum of the H_α emission corresponds to the moment when interferometer starts detecting first plasma in the vessel and shows that the line integrated density is greater than zero. This is often referred to as a delay time. An alternative approach is to define the plasma start-up time by the moment when its capable of efficiently absorbing microwaves. We identify this moment by the drop of sniffer signal in module 3 by 90% from its maximum value. The module three is used in the analysis because of its location: it is the most remote from the ECRH launchers located in modules 1 and 5. Therefore, it is not influenced by the directly reflected ECRH beam. The two different definitions are illustrated in Fig. 1. There the time traces from 02.02.2016, 15:39 UTC show (from top to bottom) the total RF power injected into the vessel, stray radiation energy flux in module 3, and line integrated density measured by interferometer. The zero moment on time axis denotes the start of ECRH operation. The left vertical bar shows the breakdown time defined by the plasma density and the right vertical bar shows the breakdown time defined by good microwave absorption.
In this contribution, we work with both of these definitions.

3 Influence of the input power and neutral gas pressure on delay time and plasma evolution

In Fig. 2 the plasma breakdown time, defined by interferometer is plotted against neutral gas pressure in module 5, measured by a manometer [μA]. The ion current is directly proportional to the neutral gas pressure. The upper panel depicts the results for helium plasmas and

[a] e-mail: dmitry.moseev@ipp.mpg.de

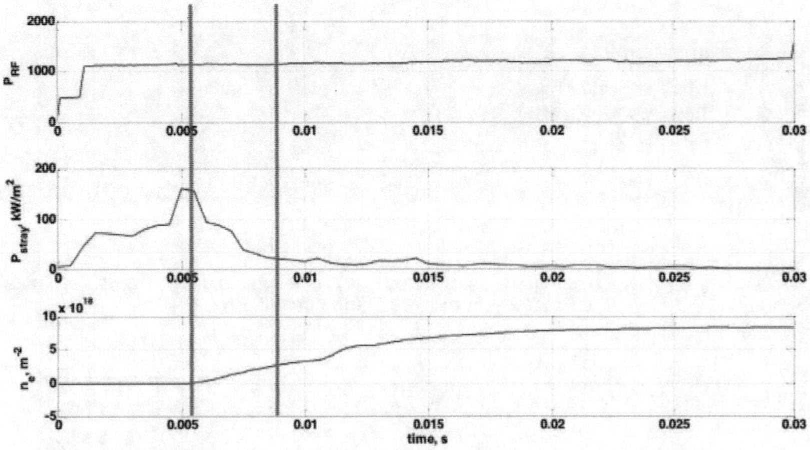

Figure 1. Definition of plasma breakdown. The example takes time traces from the experiment conducted on 02.02.2016 at 15:39 UTC. The upper panel shows total RF power injected by the ECRH system $[kW]$; the middle panel shows the energy flux of stray microwave radiation in module 3 $[kW/m^2]$; the lower panel shows line integrated density measured by interferometer $[m^{-2}]$. Zero time corresponds to the start of ECRH heating. The most left vertical bar indicates the plasma breakdown time defined by interferometer, when first plasma appears in the vessel. The most right vertical bar defines the start-up time by good microwave absorption, when the recorded stray radiation energy flux drops by 90% from its maximum value.

the lower panel shows the start-up times for hydrogen plasmas. Different power levels of ECRH are symbol- and color-coded. Counter-intuitively, the start-up times for helium plasmas are on average much longer than the corresponding start-up time for hydrogen. This is most likely due to poor conditioning of W7-X in the beginning of the OP1.1 campaign. A lot of impurities, particularly hydrogen and oxygen, were diagnosed spectroscopically in the beginning of the operational phase, and cleaning by means of the glow discharge was unavailable. The typical time traces of five experiments in helium (red) and five experiments in hydrogen plasmas are displayed in Fig. 3. One can clearly see, that the hydrogen plasma starts earlier and the state of good microwave absorption is achieved earlier as well.
Another interesting observation is that no dependence of plasma breakdown time on neutral density can be identified.

However, when the plasma start-up time is defined by good microwave absorption, the negative trend in the breakdown time versus neutral gas pressure in the launcher module is clearly visible. The corresponding graph for hydrogen plasmas is shown in Fig. 4. This behavior shows no impact of input ECRH power on the breakdown time which might indicate that the power dependence is saturated.

The dependence of density evolution on the neutral gas pressure in the launching module is positive and shown in Fig. 5. In this figure, the derivative of line integrated density is defined as $n_e^{good.abs}/(\tau_{good.abs} - \tau_{breakdown})$. In accordance with the results for the start-up time, we do not observe the dependence of the density derivative over time as a function of power, as it is shown in Fig. 6.
Previously it was shown, that the upper estimate for the start-up time in W7-X should not exceed several tens milliseconds [6]. The results from the OP1.1 experimental campaign show that the plasma breakdown needs less than the estimated upper limit.

4 Conclusions

In total more than 800 experiments were conducted in the first experimental campaign OP1.1 at W7-X. Albeit poor conditioning, the ECRH plasma start-up proved to be very reliable. The delay time was always shorter than 30 ms and typically less than 15 ms for hydrogen plasma. We did not observe the dependence of the delay time on neutral gas pressure. However, the breakdown time defined by good microwave absorption is inversely proportional to the neutral gas pressure. It should be noted that the dependence is not very strong. No significant impact of the ECRH power is seen on the start-up process. The line integrated density evolves faster when the neutral gas pressure is higher, and yet again no effect of the ECRH is visible. These all might indicate that the power effect is saturated at lower values.

Acknowledgments

This work has been carried out within the framework of the EUROfusion Consortium and has received funding from the Euratom research and training programme 2014-2018 under grant agreement No 633053. The views and opinions expressed herein do not necessarily reflect those of the European Commission.

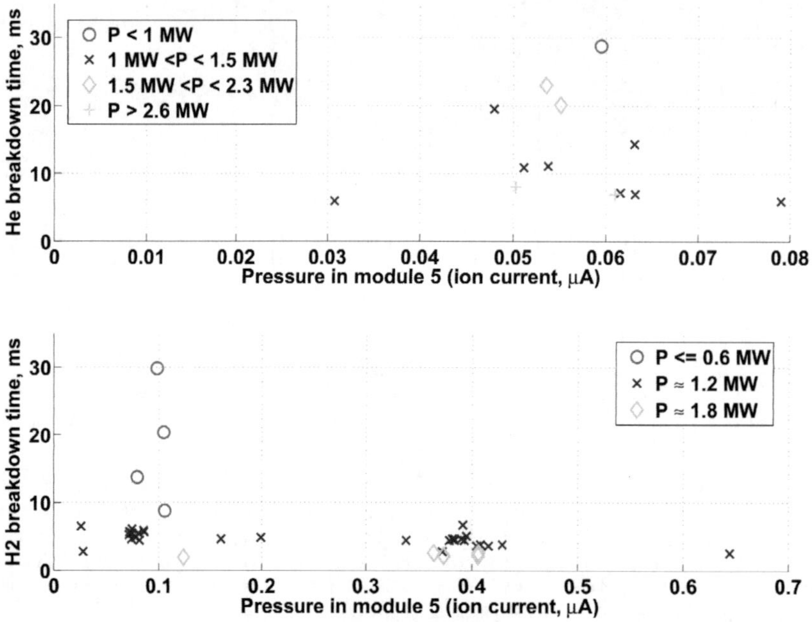

Figure 2. Plasma breakdown time, defined by the interferometer, is plotted against the manometer readings in module 5 of W7-X, where the ECRH launcher is located. The upper panel shows the start-up time for helium and the lower panel depicts start-up time for hydrogen plasmas. Power levels are symbol- and color-coded.

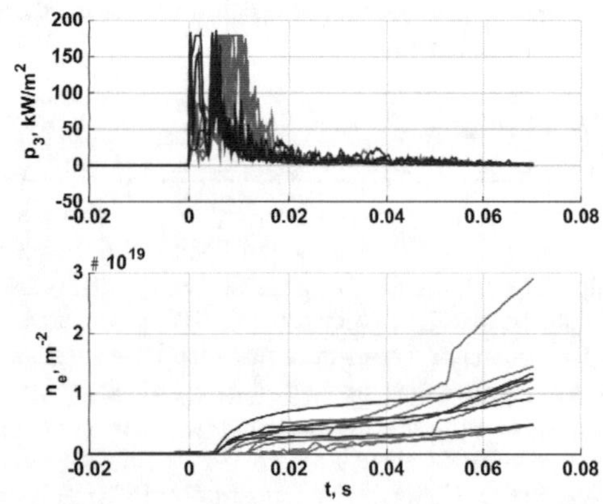

Figure 3. Time traces showing typical experiments in hydrogen plasmas (blue) and helium plasmas (red). Upper panel depicts temporal evolution of stray radiation energy flux in module 3; the lower panel shows temporally resolved line integrated density. Zero on time axis corresponds to the start of ECRH.

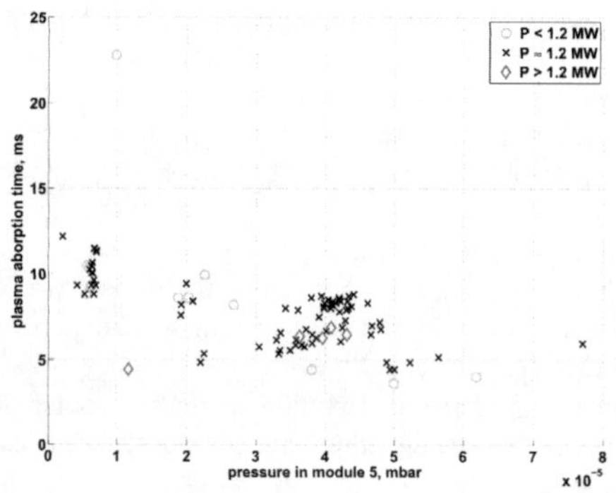

Figure 4. Plasma breakdown time in hydrogen plasmas, defined by good microwave absorption, is plotted against the manometer readings in module 5 of W7-X, where the ECRH launcher is located. Power levels are symbol- and color-coded.

References

[1] V. Erckmann, P. Brand, H. Braune, G. Dammertz, G. Gantenbein, W. Kasparek, H.P. Laqua, H. Maassberg, N.B. Marushchenko, G. Michel et al., Fusion Science and Technology **52**, 291 (2007)

[2] V. Erckmann, P. Brand, H. Braune, G. Dammertz, G. Gantenbein, W. Kasparek, H.P. Laqua, G. Michel, M. Thumm, M. Weissgerber, *The 140 GHz, 10 MW, CW ECRH Plant for W7-X: A Training Field for ITER*, in *Fusion Energy 2006* (International Atomic Energy

Agency, Chengdu, 2007), pp. IT/2–4Rd
[3] J. Geiger, R.C. Wolf, C. Beidler, A. Cardella, E. Chlechowitz, V. Erckmann, G. Gantenbein, D. Hathiramani, M. Hirsch, W. Kasparek et al., Plasma Physics and Controlled Fusion **55**, 014006 (2013)
[4] D. Moseev, H.P. Laqua, S. Marsen, T. Stange, H. Braune, V. Erckmann, F. Gellert, J.W. Oosterbeek, Review of Scientific Instruments **87**, 083505 (2016)
[5] A. Cappa, F. Castejón, F. Tabarés, D. Tafalla, Nuclear Fusion **41**, 363 (2001)
[6] M. Preynas, D. Aßmus, H. Igami, S. Kado, S. Kobayashi, S. Kubo, H. Laqua, T. Mutoh, K. Nagasaki, M. Otte et al., EPJ Web of Conferences **87**, 02005 (2015)

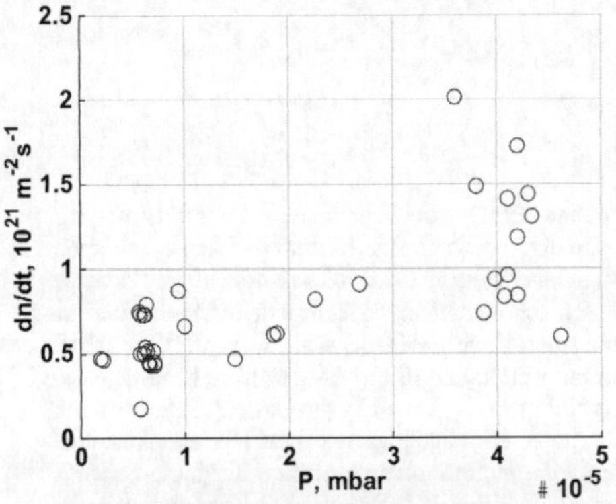

Figure 5. Derivative of the line integrated density in the period between plasma breakdown and the moment of good absorption for hydrogen plasmas plotted as a function of the average neutral gas in the launching module.

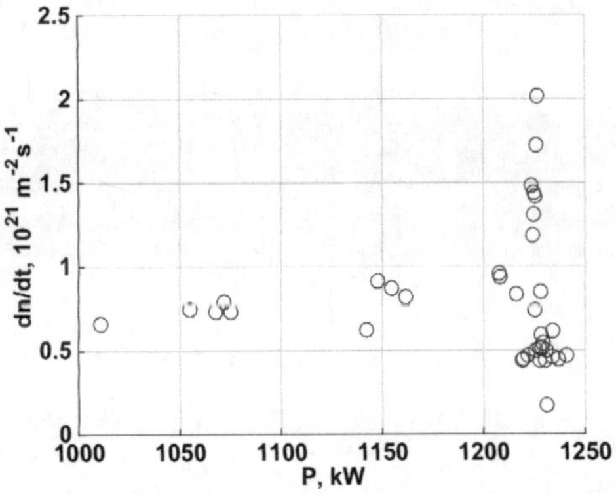

Figure 6. Derivative of the line integrated density in the period between plasma breakdown and the moment of good absorption for hydrogen plasmas plotted as a function of the ECRH power injected into the vessel.

Protecting Against Damage from Refraction of High Power Microwaves in the DIII-D Tokamak

John Lohr[1,*], *Rigo Brambila*[1], *Mirela Cengher*[1], *Xi Chen*[1], *Yuri Gorelov*[1], *William Grosnickle*[1], *Charles Moeller*[1], *Dan Ponce*[1], *Ron Prater*[1], *Antonio Torrezan*[1], *Max Austin*[2], *Edward Doyle*[3], *Xing Hu*[4], *and Calvin Dormier*[4]

[1]General Atomics, P.O. Box 85608, San Diego, California, USA
[2]University of Texas-Austin, Austin, Texas, USA
[3]University of California-Los Angeles, Los Angeles, California, USA
[4]University of California-Davis, Davis, California, USA

Abstract. Several new protective systems are being installed on the DIII D tokamak to increase the safety margins for plasma operations with injected ECH power at densities approaching cutoff. Inadvertent overdense operation has previously resulted in reflection of an rf beam back into a launcher causing extensive arcing and melt damage on one waveguide line. Damage to microwave diagnostics, which are located on the same side of the tokamak as the ECH launchers, also has occurred. Developing a reliable microwave based interlock to protect the many vulnerable systems in DIII-D has proved to be difficult. Therefore, multiple protective steps have been taken to reduce the risk of damage in the future. Among these is a density interlock generated by the plasma control system, with setpoint determined by the ECH operators based on rf beam trajectories and plasma parameters. Also installed are enhanced video monitoring of the launchers, and an ambient light monitor on each of the waveguide systems, along with a Langmuir probe at the mouth of each launcher. Versatile rf monitors, measuring forward and reflected power in addition to the mode content of the rf beams, have been installed as the last miter bends in each waveguide line. As these systems are characterized, they are being incorporated in the interlock chains, which enable the ECH injection permits. The diagnostics most susceptible to damage from the ECH waves have also been fitted with a variety of protective devices including stripline filters, thin resonant notch filters tuned to the 110 GHz injected microwave frequency, blazed grating filters and shutters. Calculations of rf beam trajectories in the plasmas are performed using the TORAY ray tracing code with input from kinetic profile diagnostics. Using these calculations, strike points for refracted beams on the vacuum vessel are calculated, which allows evaluation of the risk of damage to sensitive diagnostics and hardware.

* Corresponding author: lohr@fusion.gat.com

 DOI: 10.1051/epjconf/201714703003

1 Introduction

The Electron Cyclotron Resonance Heating (ECRH) system on the DIII-D tokamak uses gyrotron tubes generating nominally 1 MW each at 110 GHz, which, after accounting for transmission losses, inject approximately 75% of the generated power into the tokamak. The system uses launchers with flexible steering so that the rf beams can intersect the second harmonic resonance nearly anywhere in the tokamak upper half plane, with toroidally oblique angles up to about ± 20° from perpendicular for current drive. For low enough electron density, the rf propagates in nearly straight lines until being absorbed by the plasma at the second harmonic resonance. The right hand cutoff frequency is given for perpendicular incidence by

$$f_R=0.5[f_{ce}+(f_{ce}^2+4f_{pe}^2)^{0.5}],$$

which depends on the electron density through the electron plasma frequency term, where

$f_{pe}=9x10^3\ n_e^{0.5}$ Hz, and n_e is in cm^{-3}.

If the density is high enough, the right hand cutoff layer can be at a frequency higher than the second harmonic over a part of the plasma, the specifics of which depend primarily on the density profile. Injected microwaves, which encounter the cutoff prior to being absorbed at the second harmonic, will be reflected or strongly refracted with little absorption. These reflected beams, at nearly full power, can enter diagnostic ports or strike windows, causing damage to components. This possibility could be a factor affecting the ITER operating space and poses risks for diagnostics on present day machines. On DIII-D, a number of diagnostics and ancillary hardware have been damaged by refracted and reflected rf. This paper describes several of the situations which resulted in damage, and mitigating actions taken to prevent recurrence.

2 Inadvertent Overdense Operation

Overdense operation, resulting in refraction of the injected rf beams, can occur when the plasma density increases upon transition from L- to H-mode operation, when wall conditions are conducive to poor density control, with gas or pellet injection, when the poorly understood phenomenon of ECH density pumpout fails to control the density as desired, or under a number of other circumstances. When the rf fails to reach the second harmonic resonance owing to refraction, the results can be unpredictable, therefore all overdense operation is unintentional from an experimental point of view. As the density profile changes, rapid transitions from non-refractive to refractive behavior can result, with sweeping of the rf beams over a large fraction of the tokamak low field side wall, including into diagnostic ports, during the density increase. An example of extreme overdense operation is shown in Fig. 1.

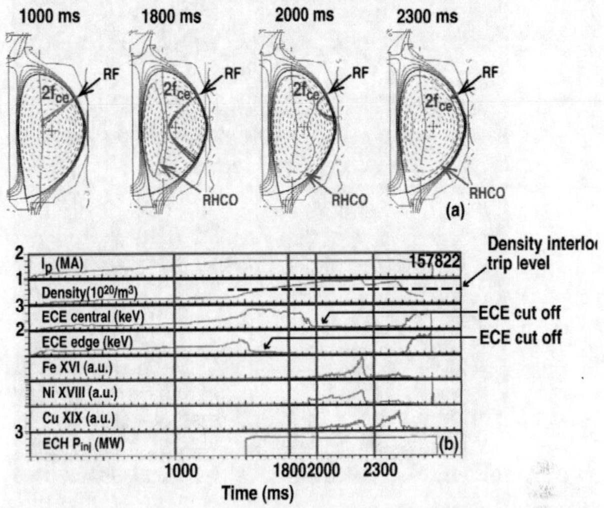

Fig. 1. This figure shows the progression of an extremely overdense plasma shot, on which the electron density by 2300 msec was high enough that the right hand cutoff frequency was immediately in front of the EC launchers, reflecting the rf power directly back into the launcher port. While the density increased, the rf beams swept across the low field side ports of the vacuum vessel. The complete cutoff of the ECE signals, a proxy for the existence of the cutoff layer, indicated that none of the injected rf power reached the second harmonic layer where it would have been absorbed. The label RHCO identifies the right hand cutoff layer. The metal lines seen spectroscopically indicate material from the ECH launchers was present in the plasma.

The ray tracing code TORAY was used to calculate the rf trajectories, using the plasma equilibrium for the shot and measured kinetic profiles normalized to a j(r) from the motional Stark effect diagnostic. In this case, the density increased until the right hand cutoff layer was directly in front of the launchers. An rf driven arc developed in one launcher, which caused melting of one of the two waveguides in the dual waveguide configuration. A Langmuir probe mounted at the output end of the waveguide was vaporized and the insulating ceramic beads on the bias wire were melted into glass. Because of the damage to the wires, the ECH permit was not removed by the Langmuir probe interlock. The damage is shown in Fig. 2.

 DOI: 10.1051/epjconf/201714703003

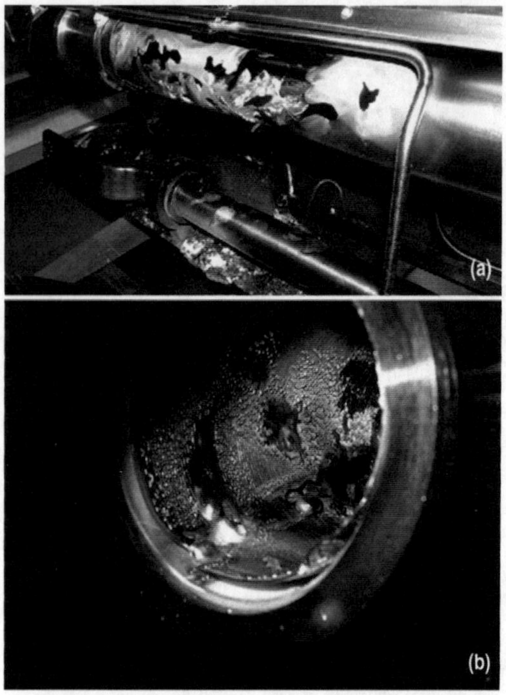

Fig. 2. The shot documented in Fig. 2 resulted in damage to one of the EC launchers, in which extensive melting of the waveguide and wires occurred. Fig (a) shows a view of the melted waveguide, and (b) is a photograph of the launch end of the waveguide. The damage was likely caused by rf heating of an arc, possibly where the third harmonic electron cyclotron resonance was present.

As can be seen from the progression of the beam trajectories in Fig. 1, any diagnostic on the low field side of the tokamak in the ECH launcher sector could have been subjected to rf heating as the beams swept across the ports. Microwave diagnostics, such as the ECE imaging diagnostic, located in the midplane port below the EC launchers, had suffered damage in the past and viewports had also been cracked, resulting in partial venting of the DIII-D vacuum vessel. It became clear that additional measures would be required to prevent damage to the installation under overdense conditions.

3 Mitigation

Mitigation of the risk of damage from ECH injection into the DIII-D tokamak has taken several forms. Administrative measures were developed, diagnostics were fitted with protective filters, and procedures were established to determine whether, given the operational parameters for the tokamak and the ECH system, there was a risk of damage. Additional diagnostics to provide risk assessment and protective fast response were installed.

A set of procedures was developed to clarify the allowable operational parameters. A density interlock was created based on the best available measurements as determined by the plasma control system. Each morning, this density limit was set automatically to a line average of $5.6x10^{19}/m^3$. This interlock line is shown on the time traces in Fig 1 b. The first shot of the day, or the first shot after a major configuration change, would be required to be without ECH. Using the kinetic profiles from this shot, the ECH operators then calculate the rf beam trajectories using TORAY for the equilibrium and launcher aiming. Based on these calculations, the default density trip point could be changed to accommodate the experiment at as high a density as would be safe for the machine and diagnostics. A feature was added to the ray tracing code that calculates the trajectory of the rf rays for cases where >10% of the injected rf power fails to be absorbed for any of the gyrotrons. This allows a determination of the likelihood that a particular set of plasma parameters and trajectories will result in damage, by identifying the ports into which power might be deposited. This calculation is shown in Fig. 3, for the shot in Fig. 1.

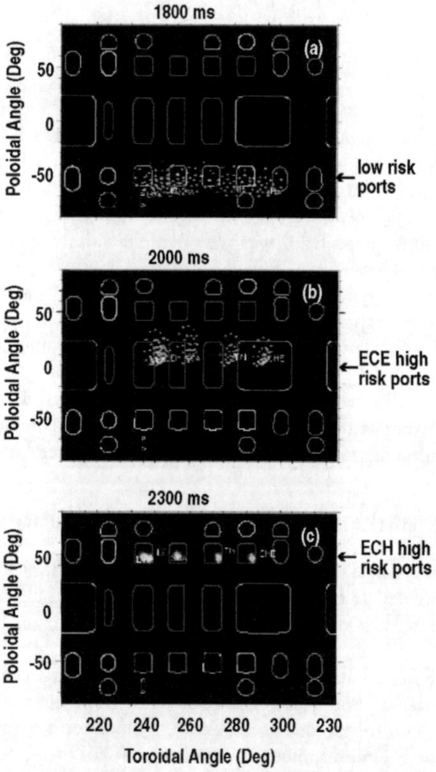

Fig. 3. In order to help assess the risk to the tokamak from rf refracted from the right hand cutoff layer, a code was written as an adjunct to TORAY, which calculates the points at which the refracted rf beams strike the vessel wall on the low field side. The plots show which ports will be struck for the plasma equilibrium and ECH launcher mirror steering in use using a multi-ray approximation to a free space Gaussian rf beam. Each rf beam is modeled by 100 rays. The strike points for each ray are plotted.

In order to support these procedures, the accuracy of the diagnostic data, particularly the density profile, with Thomson scattering data verified by ECE cutoff and reflectometer data, and with equilibria normalized to j(r) measured by the Motional Stark Effect diagnostic, was improved. Although the operating rules allowed a 10% safety factor for the density trip point to be set, with improved accuracy of the data used in the trajectory calculation, this safety factor cold be more accurately set to maximize the allowable parameter ranges.

3.2 Protective Hardware

3.2.1 Sniffer probes

At several fusion research facilities attempts have been made to develop a detection system for stray rf power, which could serve as the basis for a protective interlock [1, 2]. At DIII-D, efforts were made to develop an rf detector, or "sniffer probe", which could be mounted at several locations around the tokamak vacuum vessel to help diagnose incomplete absorption of the injected rf and the potential for damage to occupants of ports. These tests did indicate when excessive rf was reflecting off the vessel walls, but showed that using the signals as a protective interlock resulted in an unacceptably large number of false trips, since we could not compensate for the polarization sensitivity of the detectors and the large signal variation arising from the angles with which the rf beams propagated in the vessel. One proposed solution to this problem has been to time-integrate the stray rf signal with a large cavity in which polarizations and trajectories would be isotropized prior to detection. These spheres worked in tests, but they occupied too much real estate in the close confines of the tokamak environment [3], at least on DIII-D. A new design for a sniffer probe has recently been presented [4] and this bolometer based probe is being evaluated on DIII-D.

Even if sniffers, which have a typical response time of 100 msec for bolometers, can alert regarding low absorption of ECH, they cannot respond quickly enough to protect diagnostics based on mm microwave technology employing diode detectors and mixers. Therefore protective measures must be installed on the diagnostics themselves. These steps have been taken on diagnostics systems on DIII-D.

3.2.2 Filters

The safest and most reliable protection against damage from ECH for diagnostics is for the diagnostics to be fitted with filters that reject the 110 GHz ECH microwave frequency. For receivers that use waveguide transmission lines, stripline filters can be used [5]. These are shaped conductors incorporated into a waveguide carrying the signal. The conductors are separated from a ground plane by a dielectric and the assembly is mounted in a housing connected to the waveguide. Depending on the shape of the conductors, bandpass, band reject, high or low pass filters can be designed using striplines. For microwaves in the 100 GHz range, high precision is required in the geometry of the structures, both in shape and thickness of the conductors. Stripline notch filters have been installed on the microwave reflectometers at DIII-D and provide about -30 dB rejection.

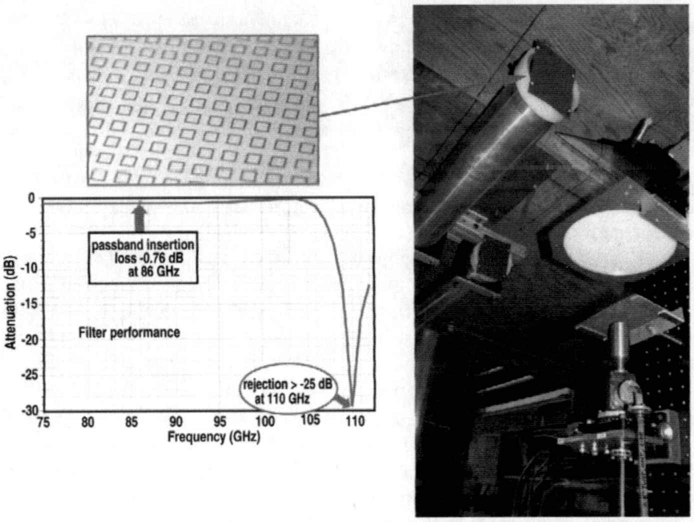

Fig. 4. Planar resonant cutoff filters produce a -30dB notch at the 110 GHz gyrotron frequency, which can be increased to -60dB attenuation by stacking three of the filters in series. Insertion loss off-resonance is < -1 dB. The filters are produced by deposition of the repeated resonant structures onto a dielectric substrate using precise photolithographic techniques. The filters are easy to install and can be cut to shape using simple tools.

A second class of notch filters uses an array of resonant structures applied to a dielectric substrate using printed circuit technology [6,7]. Although these filters, Fig. 4, are continuing to be improved, they can produce a notch width of about 5 GHz at -3dB, an insertion loss of less than 1 dB over a wide band, and a notch depth of -30 dB. By stacking up to 3 filter sheets in series with a few degree relative tilt, notch depths up to -60 dB have been achieved. These can be combined with a dichroic plate providing high pass filtering to protect large arrays of microwave mixers and detectors used for imaging the plasma in the electron cyclotron radiation emitted from resonant surfaces, commonly the second harmonic of the electron cyclotron frequency. The filter sheets can be tilted up to 15 degrees without significantly affecting the transmission, making possible the series stacking mentioned above without developing standing waves due to reflections. The sheets can be cut with shears. In some cases both stripline and quasi-optical filters are used in series, providing about -40 dB rejection.

3.2.3 Light monitors

A basic diagnostic, which has been in use on DIII-D for many years, is simple video imaging of the structure of the rf launchers seen through windows mounted in the launcher flange. The windows typically are sapphire in order to avoid microwave absorption, and provide a relatively good view of the launcher hardware through a wide angle lens. The video signals are observed by the ECH Operators, looking for excessive brightness or unexpected light persisting for seconds after the plasma shot, indicating incandescence from launcher components at high temperatures. The video images are archived on a multi-channel recorder allowing unusual events to be examined in more detail after the shot. In Fig. 5, an example of two video frames is shown, the first of which was taken during the ECH part of a plasma shot and the second, which showed persistent light about 5 sec after the end of the plasma current from incandescence of melting launcher parts. In this case, the video monitors alone alerted the operators of a problem in the launcher.

In order to reduce the requirement for operator intervention, each waveguide line on the DIII-D system has been equipped with a visible light monitor to function in concert with the video system. The monitors are compact and mount directly on a sapphire viewing port with about the same field of view as the video cameras. Each light detector produces a modulated output, the frequency of which is proportional to the brightness of the detected light. This signal is routed to signal processing boards, which use f-V chips to produce an analog signal proportional to the detected light, but which is relatively immune to isolated light bursts and electrical noise. These signals are incorporated into an alarm system, which looks for light after the end of the plasma current.

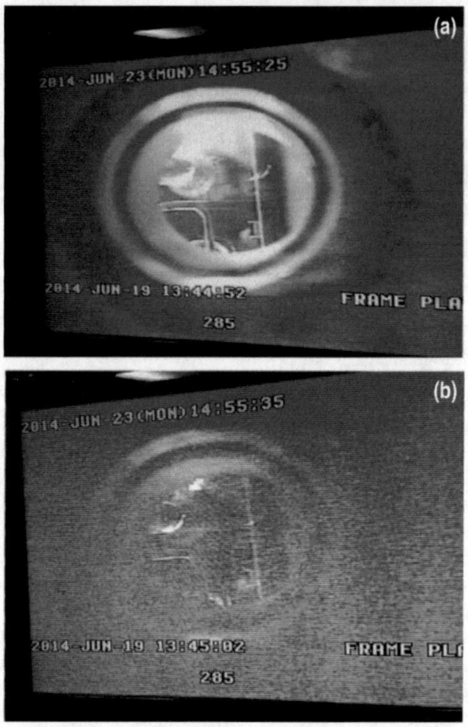

Fig. 5. These video images taken looking through windows in the EC launcher flanges provide a view of the launchers both in plasma light and in light from incandescent launcher assemblies long after the plasma current has ended. View (a) was taken during the plasma shot, and (b) was about 5 sec after the end of the plasma, indicating the brightness of the incandescence well after the shot. Rather than rely on the ECH operators to take note of the persistent light following a shot, a set of visible light detectors were installed, which were interlocked to provide an alarm if the detected light persisted following the end of a plasma shot.

3.2.4 Langmuir probes

Presence of plasma in the vicinity of the launcher waveguides can be detected by Langmuir probes which are installed to the side of each waveguide mouth FIG. Each of these provides an interlock signal, which removes the ECH permit from the gyrotron connected to that waveguide. In some cases, ELMS, disruptions, pellet injection or instabilities can trip the Langmuir probe interlocks. In the highly overdense case described here, a flaw in the DIII-D launcher probe design was discovered. When the rf beams were reflected directly back into the launchers, the bias lead was immediately vaporized on one probe, melting the ceramic insulating beads and eliminating the probe signal being fed to the outside. The rf driven arc then caused the extensive waveguide melt damage seen in the photographs. To counteract this in the future, probe bias feed wires have all been placed in metal shields. It is hoped that now the probes will terminate the rf pulses prior to onset of damage.

3.2.5 Microwave detectors

On all waveguide lines in the DIII-D system, the final miter bend in each waveguide line has now been instrumented to provide additional information, which can be used to interlock the rf generation. The advanced miter bends have been converted to 4 port devices with perforated mirrors that reflect most of the rf but pass a polarization insensitive signal at the -75 dB level, which can be used with an orthomode transducer to measure the linear polarization of the rf beam. Using mode sensitive directional couplers, the miters can measure the content of several low order modes, $HE_{1,1}$, $HE_{2,1}$, or $TE_{0,1}$, for example. The miter arms also can detect forward and reflected rf power, allowing an interlock to be based on the forward/reflected power ratio. These devices are now installed on all the ECH waveguide lines and are being set up to provide the interlock functions.

4 Conclusions

High power electron cyclotron resonant microwaves can be generated and directed with good precision into tokamak plasmas, heating plasma locally and driving current. Because the plasma density can be high enough that refraction and reflection can occur before the rf beams reach the resonance, where essentially complete absorption will occur, instead the rf rays can follow a curved path so the power can be absorbed at unintended flux surfaces intersecting with the rays and the resonance. If the refraction is strong, the rf can easily be reflected to unintended parts of the vacuum chamber, where it can be absorbed by diagnostics or other objects such as viewports, causing damage, venting of the vacuum vessel or even damage to the ECH system itself.

On DIII-D, after several cases of damage had occurred, mitigating measures were implemented. These included administrative procedures, installation of protection such as filters on the vulnerable diagnostics and equipment, and sensors on the ECH launchers to alert when densities have reached dangerous values. Additional measures are being considered, but in the meantime, no additional cases of damage have occurred anywhere on the tokamak from the ECH system.

References

1. F. Gandini, *et.al.*,Fusion Eng. Des., **56-57,** 975 (2001).

2. H. Igami, *et.al.*,Searching for O-X-B mode-conversion window with monitoring of stray microwave radiation in LHD, Rev. Sci. Instruments, **77,** 10E931 (2006).

3. M. Schubert, *et.al.*, Monitoring millimeter wave stray radiation during ECRH operation at ASDEX-Upgrade, *Proc. 17th Joint Workshop on Electron Cyclotron Emission and Electron Cyclotron Resonance Heating, EPJ Web of Conferences* **32**, 02013 (2012).

4. H. Oosterbeek, Design and R&D for the ITER Stray Radiation Bolometers, *Proc. 19th Joint Workshop on Electron Cyclotron Emission and Electron Cyclotron Resonance Heating, Institute for Plasma Research, India* (2016).

5. Terry Edwards and Michael Steer, Foundations for Microstrip Circuit Design, John Wiley and Sons, West Sussex, UK, ISBN:9781118936191 (2016).

6. H.J. van der Meiden, Rev. Sci. Instrum., **70**, 6, 2861, (1999).

7. Z. Shen, *et.al.*, Plasma and Fusion Research, **2**, S1030 (2007).

This work is based upon work supported by the US Department of Energy, Office of Science, Office of Fusion Energy Sciences, using the DIII-D National Fusion Facility, a DOE Office of Science user facility under awards DE-FC02-04ER54698[1], DE-FG03-97ER54415[2], DE-FG02-08ER54984[3], and DE-FG02-99ER54531[4].

Summary of papers on technology

Burkhard Plaum[1,*]

[1]Institut für Grenzflächenverfahrenstechnik und Plasmatechnologie, Universität Stuttgart, D-70569 Stuttgart, Germany

Abstract. The contributions on technology are summarized.

1 Gyrotrons

Gyrotron development was with 15 submitted abstracts the largest technology topic on the workshop, which shows that the microwave tubes are still the most challenging components in all EC heating systems.

The development status of the Russian ITER gyrotron was reported by D. Denisov. The Russian tube produces 0.96 MW at the MOU output at ITER relevant pulse lengths (up to 3600 s). Another series of gyrotrons developed by the Russian Gycom company for frequencies in the range of 105 GHz to 140 GHz was presented by E.M. Tai.

The European ITER gyrotron status was reported by G. Gantenbein. After successful tests with a short pulse tube, a 1MW CW prototype was built and tests started in the beginning of 2016.

J. Jelonnek presented the design status for a European 240 GHz coaxial cavity gyrotron for Demo. Several topics, like frequency step tunability and a multistage depressed collector were discussed.

A proposal of mode selection criteria for a 255 GHz gyrotron for the Korea demonstration power plant (KDEMO) was discussed by A. Sawant. The power will be > 1 MW with a conventional cavity operating at the $TE_{55,13}$ mode.

Numerous presentations came from IPR colleagues about the domestic gyrotron development in India. NK. Singh gave an overview of the 42 GHz, 200 kW Gyrotron, which is developed by IPR for use at the Tokamaks SST1 and ADITYA. The development of an anode modulator power supply (AMPS) with a crowbar protection system for 82.6 and 42 GHz gyrotrons was reported by N. Rajanbabu.

An overview of the ITER-India EC RF source package was given by S.L. Rao. It consists of the delivery of two 1 MW/3600 s gyrotron tubes operating at 170 GHZ. It also includes the development of a Gyrotron test facility (IIGTF).

The IIGTF was presented by V. Rathod. It includes power supplies, a crowbar protection system, a dummy load, beam diagnostic systems and cooling systems.

S. Dilip reported on the prototype development of the body power supply for the IIGTF. It has the capability to modulate the cathode and body potentials simultaneously and will be realized using cost-effective high voltage solid state switches. A. Sharma presented the RF Beam diagnostics for the IIGTF. It uses thermography with an IR camera and a software module to retrieve the phase of the output beam. These data are used to characterize the output beam of the gyrotron in order to minimize the spurious modes in the transmission line.

An overview of the local control unit for the IIGTF was presented by R. Shah. The prototype of the LCU will have programmable controllers, which will allow the gyrotron operation through a graphical user interface (GUI). In the final version, the software will be compatible with the ITER CODAC core system (CCS).

In conclusion of the gyrotron related contributions, it can be said, that one problem is the fact, that there is no series production of fusion relevant high-power/high-frequency gyrotrons. This is mostly because the frequencies (which are defined by the magnetic field), are always increasing. This implies the need for higher order cavity modes, which have increased requirements for the internal mode converter and might be sensitive to mode competition.

Also, the supporting technologies like power supplies, crowbars and control systems need to be constantly adapted for the changing requirements. Thus, they are often developed in-house by the fusion research institutes.

One suggestion to work towards a series production and better competition among manufacturers would be to standardize the parameters, most notably the frequencies of gyrotrons. This would imply non-optimal frequencies in some cases (along with decreased heating efficiency) while the advantage would, in the long term, be reduced prices and improved reliability of the tubes. Considering the current experiments, one can already see a tendency into this direction, since 105, 140 and 170 GHz are already used by more than one fusion experiment.

Another topic is frequency tunability. Challenges here are the design of broadband windows (e.g. Brewster windows) and the dependence of the output power on the frequency. Finally, for Demo it will be necessary to increase the efficiency beyond the 50%, which are typically achieved today. Thus, advanced depressed collector configurations will remain an important issue.

* Corresponding author: plaum@igvp.uni-stuttgart.de

2 Launchers

Another important topic was the development of ECRH launchers. There are two competing technologies: A front steering launcher has movable mirrors close to the plasma. One example is the upper ECRH launcher for ITER, which was presented by D. Strauß. It is designed by a European consortium and will allow to use 8 beamlines with 1 MW each for MHD control. In total, 4 upper launchers are foreseen. It was stated, that the main design issues are mostly on the administrational level from the ITER side, which makes the launcher design a *moving target*. Another presentation was given by H. Mistry about an ECRH launcher for the SST-1 tokamak at IPR, which operates at 42 and 82.6 GHz and has a steering range of ±20°.

The other option is the remote steering launcher, which uses a waveguide with imaging properties and has the steering mechanism several meters away from the plasma. One remote steering antenna is developed for Wendelstein 7-X and was presented by C. Lechte.

While remote steering launchers have clear advantages, especially in reactor-like conditions or if space is limited, the disadvantage is the limited steering range.

3 Complete systems

An overview of the conceptual design of the Demo EC system was presented by S. Garavaglia. The Demo reactor will, in comparison with ITER, require higher power at higher frequencies. The requirements regarding the flexibility are, however, less strict since Demo will be a fusion reactor prototype. Starting with the baseline design "EU DEMO1 2015", the requirements for the EC system are a power of 50 MW at 170 and 204 GHz. The power per tube is 2 MW and the reference efficiency is 60%. For the launcher, only the remote steering type (see Section 2) and the open ended waveguide (without any steering) were discussed. Calculations of the current drive efficiency with the TORBEAM code were discussed as well as the impact on the tritium breeding ratio. For the transmission line, an evacuated quasi-optical line was suggested and different configurations were investigated with respect to the system reliability.

J.H. Jeong presented the development status and plan of the KSTAR ECH system in Korea. In addition to the systems running at 84 and 110 GHz (short pulse) and 170 GHz (cw), a new system, which will deliver 3 MW at 105 and 140 GHz will be installed. The presentation included an overview as well as a description of the power supplies.

4 Other Topics

The issue of stray radiation in plasma vessels was discussed by J. Osterbeek. The radiation consists mainly of non-absorbed ECRH beams, which become an isotropic radiation background after many wall reflections. The second source is the synchrotron radiation of the plasma itself, which becomes important at high electron temperatures. The unwanted effects of stray radiation are heating of components and potential damage of sensitive diagnostic elements. After an introduction to the topic, the bolometers, which will be used as stray radiation monitors in ITER, were presented.

The design of external mode converters for the Indian 42 GHz gyrotron was presented by K. Sathyanarayana. The gyrotron outputs a TE_{03} mode, while at the output of the converter chain, the HE_{11} mode is required. First, the TE_{03} mode is converted to a TE_{02} and then to a TE_{01} mode. These converters change only the radial indices of the modes, and are therefore rotational symmetric structures with a varying diameter. The TE_{01} mode in the smooth wall waveguide is degenerate with the TM_{11} mode, and the azimuthal indices differ by one. This means, that the TE_{01} mode will be fully converted to the TM_{11} mode in a bend with a constant curvature after certain angle. The $TM_{11} \rightarrow HE_{11}$ conversion is done by increasing the corrugation depth continuously from zero to $\lambda/4$. While such oversized converters have been designed for many years, they required dedicated computer codes, which calculated the mode conversion. Full-wave solvers could not be used for 3D structures, which have dimensions of many wavelengths. However, today it is possible to design such oversized components with a commercial general purpose tool.

Diagnostic components for ECRH systems also need to be adapted to the changing requirements, most notably higher power, longer pulse lengths and two (or more) frequency operation. B. Plaum presented a number of different developments from this area. Different coupler techniques, which are necessary for power monitoring, were extended for operation at 105 and 140 GHz in ASDEX-Upgrade. Grating couplers for quasi-optical systems were discussed as well as hole couplers, which can be integrated into mitre bends of waveguide transmission lines. Special arrangements of hole couplers can, together with a suitable signal analysis, used as simple detectors for spurious modes. Also, a 1 MW calorimetric dummy load was presented, which works reliably under atmospheric pressure.

The experimental verification of Gaussian beam coupling into HE_{11} transmission lines at 400 GHz was reported by M.S. Choe. Although the coupling of Gaussian beams into waveguides is established technology, the significantly higher frequency implies stricter requirements for the machining accuracy and the measurement equipment. The measured results are in good agreement with Surf3D simulations, which were used to design the coupling mirror.

A report on the instrumentation and control system architecture for the ECRH system for the tokamak SST-1 was given by H. Patel. The developments include a two level interlock system with fail-safe logic where slow signals in millisecond scales are processed by software, while hardware interlocks operate on microsecond timescales.

Progress in conceptual design of EU DEMO EC system

Saul Garavaglia[1,*], *Alex* Bruschi[1], *Thomas* Franke[2,3], *Gustavo* Granucci[1], *Giovanni* Grossetti[4], *John* Jelonnek[5], *Alessandro* Moro[1], *Emanuele* Poli[2], *Natale* Rispoli[1], *Dirk* Strauss[4] and *Quang Minh* Tran[6]

[1]Institute of Plasma Physics "P.Caldirola", National Research Council of Italy, Milan, Italy
[2]Max-Planck-Institut für Plasmaphysik, Boltzmannstr. 2, Garching, Germany
[3]EUROfusion Consortium, Boltzmannstr. 2, D-85748 Garching, Germany
[4]KIT, IAM-AWP, Postfach 3640, D-76021 Karlsruhe, Germany
[5]KIT, IHM, Postfach 3640, D-76021 Karlsruhe, Germany
[6]Swiss Plasma Center, EPFL, CH-1015 Lausanne, Switzerland

Abstract. Since 2014 under the umbrella of EUROfusion Consortium the Work Package Heating and Current Drive (WPHCD) is performing the engineering design and R&D for the electron cyclotron (EC), ion cyclotron and neutral beam systems of the future fusion power plant DEMO. This presentation covers the activities performed in the last two years on the EC system conceptual design, as part of the WPHCD, focusing on launchers, transmission lines, system reliability and architecture.

1 Introduction, physical requirements and design guidelines

According to the European Roadmap [1] the main purposes of the Demonstration Fusion Power Plant (DEMO) are to deliver electricity to the grid and to allow Tritium self-sufficient breeding in order to assure autonomy for its own operations. The EUROfusion consortium is conducting detailed studies on different aspects of DEMO power plant and the Work Package Heating and Current Drive (WPHCD) is performing conceptual studies and the engineering design for the electron cyclotron (EC) [2], ion cyclotron and neutral beam heating systems and R&D for gyrotron source and NB injector. The primary objective is to deliver a feasible concept design of the EC system fully integrated in the machine design, satisfying the stringent safety and Remote Maintenance (RM) criteria and minimizing the impact on the Tritium Breeding Ratio (TBR). Various options for gyrotrons, Transmission Lines (TL) and launchers are under assessment taking into account the integration in a nuclear environment and target RAMI (Reliability, Availability, Maintainability and Inspectability) requirements specific for a power plant.
On the one hand, EU DEMO1-2015 is the present baseline design, a 2 h pulsed machine, with an aspect ratio (AR) of 3.1, toroidal magnetic field of 5.7 T and about 50 MW external heating and a minor role of auxiliary plasma current drive (CD). On the other hand, EU DEMO2-2015 is a steady state machine (i.e. a more advanced concept) where CD is significant to sustain the plasma current. Table 1 summarizes the main basic tokamak parameters derived from the PROCESS code [3]. The starting point of the conceptual design is the identification of the physical requirements demanded to the EC system for EU DEMO1-2015.

Table 1. Summary of major tokamak parameters for DEMO1 and DEMO2 (April 2015).

	EU DEMO1 2015	EU DEMO2 2015
Major radius [m]	9.072	7.5
Minor radius [m]	2.927	2.885
Aspect Ratio	3.1	2.6
Toroidal field [T]	5.7	5.627
Plasma current [MA]	19.6	21.6
Heating power [MW]	50-100	133
Fusion Power [MW]	2037	3255
q_{95}	3.247	4.405
Number of TF coils	18	18
Pulse duration [hours]	2	continuous
$<n_e>[10^{20}\ m^{-3}]$	0.8	0.9
Peaking: $n_{e0}/<n_{e,vol}>$	1.27	1.397
$<Te>$ [keV]	13.1	18.1
Peaking: $T_{e0}/<T_e>$	2.1	1.9
Surface area [m^2]	1428	1253
Plasma volume [m^3]	2502	3255
Fraction of $I_{BOOTSTRAP}$	0.347	0.611
Fraction of I_{OHMIC}	0.557	0
Fraction of I_{CD}	0.096	0.389

The main EC tasks are reported in Table 2 with the required power and deposition localization in terms of normalized radius. The preliminary studies suggest that

* Corresponding author: garavaglia@ifp.cnr.it

4-6 MW are needed to sustain the DEMO breakdown and start-up and point out that an oblique injection is desirable to exploit the polarization conversion of non-absorbed power at inner wall. For plasma current ramp-up preliminary results with the METIS code [4] indicate 50 to 100 MW of additional power are needed to reach robust L-H transition. During plasma flat top presently 50 MW of HCD power is considered, assuming in this paper that EC providing the full amount of this power, with both oblique or perpendicular injecting angle. The later DEMO HCD mix selection may change these EC power requirements. During the burning plasma the EC system is also required for MHD control: 10-15 MW of EC power is dedicated for Neoclassical Tearing Mode (NTM) and sawtooth control. For the safe plasma current ramp-down up to 40 MW of off-axis EC power is considered by the simulations with JINTRAC code [5] to avoid edge cooling and widen the current profile keeping plasma internal inductance lower.

Table 2. Main DEMO EC tasks with corresponding power required and plasma deposition location.

		Power required [MW]	Localization [ρ]	Mode
Assisted Breakdown		6-10	0-0.3	Heating /CD
Ramp up and L-H transition		50-100	0-0.3	Heating /CD
Main heating		50	0-0.3	Heating /CD
NTM control	q=2	10-15	0.7	CD
	q=3/2	10	0.4	CD
Ramp down		40	0.3	Heating

A total of 50 MW of EC power in addition to 10 to 15 MW EC power dedicated to NTM control in the plasma has been taken as reference for the present analysis; this total EC power must be guaranteed at maximal reliability and availability for all the DEMO pulses because an interrupted discharge will mean no electricity production.

The EC power will come from gyrotrons sources with estimated 2 MW output power per RF source, an efficiency of 60 % as target (cf. ITER 50 % [6]) and 98 % of unit reliability. Multi-purpose (multi-frequency) and frequency step-tunable gyrotrons are under investigation to fit all the physical requirements of Table 2. Multi-purpose aims at the gyrotron operations with different magnetic field configurations (slowly varying) and different possible frequencies for heating and CD corresponding to multiples of the λ/2 wavelength of the RF diamond window (~34 GHz for typical single-disc window thickness of ~1.8 mm), or, alternatively a broadband window design. The present reference frequencies are 170/204 GHz while the final operating ones will depend on the AR definition and toroidal related magnetic field and on the relevance of CD in operation, which typically requires an operation at about 1.2 times higher frequency compared to pure heating. On the other hand the availability of frequency step-tunable gyrotrons using a broadband RF output window (e. g. Brewster-angle window) which will enable the operation in steps of about 2 to 3 GHz (according to the different frequency distances of the operating modes) over an 10 to 12 GHz bandwidth will be compatible with the exploitation of a remote steering antenna concept for the launchers as presented at the EC-19 conference [7]. Equatorial (EL) and vertical (VL) launchers (without the use of switch between them) are required to deliver the RF power to the plasma through apertures into the blanket, that shall minimize the impact on Tritium Breeding Ratio (TBR). The EC system design must be as much simplified as possible with a single purpose demanded to each EC line, compatible with Remote Maintenance and involving the modularization of components to pursue economic improvement.

2 Launchers

A launcher with a sufficient flexibility and without movable parts in the proximity of plasma and blanket is required to deliver the required amount of power at different deposition locations. A Remote Steering Antenna (RSA), able to grant a continuous but limited steering range without mirrors in plasma proximity and Truncated Waveguide (TWG) launching a divergent Gaussian beam in a direction determined by waveguide orientation have been considered so far as launcher solutions. A general assessment based on previous DEMO baseline design (EU DEMO1 2012 with AR = 4.0) of the RSA capability has been performed in terms of launching performance, plasma accessibility, RS properties and potentialities for multi-frequency gyrotrons.

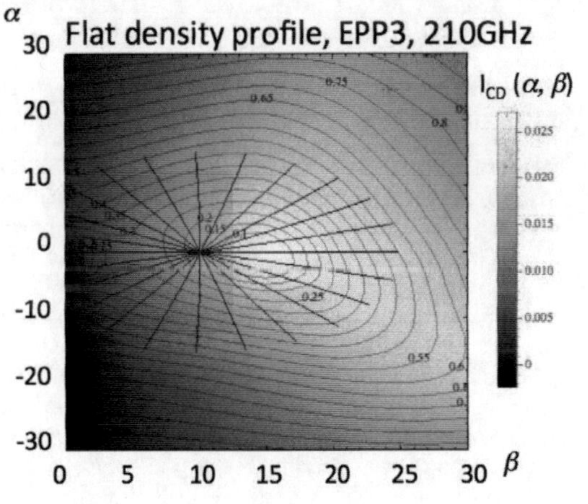

Fig. 1. Contour plots for normalized deposition location ρ and total driven current I_{CD} (MA/MW) as a function of the injection angles (α, β). The case of 210GHz frequency is shown, with launch point EPP3, the set of steering planes investigated (blue lines) and the port geometric constraints (shadowed area in green).

A possible integration into port plug and preliminary evaluation of required apertures for the RSA assembly was started and will be completed next year. In order to perform the analysis on possible launching configurations the beam tracing code TORBEAM [8] was used to simulate EC injection in several configurations. Five different launching points have been considered from both Equatorial (EPP) and Vertical Port Plug (VPP). The input parameters for TORBEAM runs are frequency f (170 ÷ 250 GHz), launched beam dimensions w_0 = 20.43 mm, toroidal angle β (0° ÷ 40°) and poloidal angle α (-45° ÷ 45° for EPP and 30° ÷ 60° for VPP) for two plasma profiles with density and temperature flat and peaked respectively. For a given scenario, frequency and launching point an optimal pair of reference injection angles (α_0, β_0) can be found and used to identify possible RSA planes and maximum steering ranges of application. As an example of the performed analysis Figure 1 shows the contour plot of normalized deposition location ρ and total driven current I_{CD} as a function of (α, β) angles□ All the results of beam tracing analysis are summarized in a map of I_{CD} for selected RSA direction and steering plane.

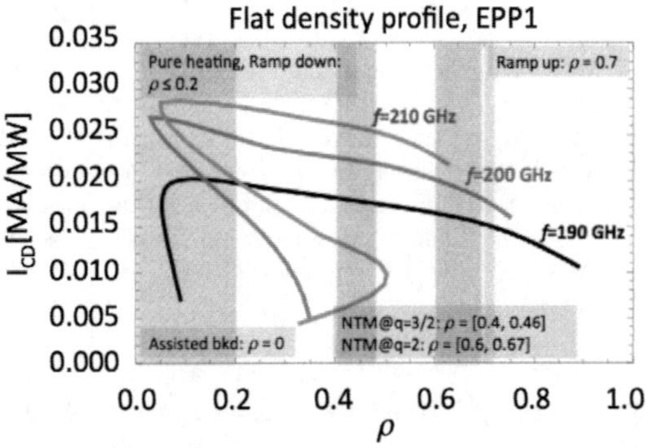

Fig. 2. Summary of total driven current I_{CD} (MA/MW) as a function of accessible deposition location ρ with the beam launched from EPP1 with α_0=-10°, β_0=17°, 6□□□□° for frequency in the range [190-210GHz]. Required EC functions are shown with rectangular areas.

In Figure 2 the most promising solution from EPP is shown where a complete coverage of deposition locations of EC functions is reached for f in the range 190 ÷ 200 GHz. Vertical port results show a better CD efficiency but reduced plasma coverage. A preliminary estimation of the minimum apertures required by the launchers on the Breeding Blanket (BB) has been calculated in order to allow a first evaluation of the impact on TBR of the EC launchers. A straight square corrugated WG has been considered as RSA termination with 63.5 mm size, steering range ±15 degrees, not protruding in the blanket region with 8 beams in two rows for EL and 4 beams for VL. The EC frequency assumed is 170 GHz and a Gaussian beam with 20 mm waist at the waveguide output. For each antenna, a space of 15x15 cm is foreseen to accommodate components, cooling and supporting structures and any auxiliary equipment needed at the port interior. The openings at BB level resulting are ~2.34 m^2 for 4 (+1 spare) ELs and 1.24 m^2 for 2 VLs (including 3 spare lines).

Tritium self-sufficiency is mandatory for DEMO and a net value of TBR ≥ 1.1 has to be reached including an additional margin, which account for modelling uncertainties and plant losses occurring during DEMO operations. Starting with the apertures calculated at the blanket level, the impact on the TBR for the five ELs is ΔTBR of ~0.004 (~0.313 %) and for the two VLs ~0.002 (0.166 %). These values look promising to reach the target value of 1.1 and with safety margins.

3 Transmission Line

On present experimental fusion devices two solutions for TL are adopted: Evacuated waveguide (EWG) in experiments as e.g. DIII-D, TCV and Quasi-Optical (QO) in air as e.g. used for W7-X. The main DEMO TL requirements are: target efficiency of 90 %, power handling of 2MW CW, multi-frequency (or broadband) capability and tritium compatibility. EWG is certainly contemplated a possible solution for DEMO but, since the WG components are still under development for ITER, the QO solution based on recent W7-X experience has been first considered. For DEMO the transmission of 10 beams in 2 Multi-Beam TL (MBTL) could be a very compact arrangement to reduce the complexity of the system and to save space and components. This solution is promising for the large number of DEMO beams provided that the distance is not excessively long. However the power transmission in air is not compatible with a nuclear plant for tritium segregation (in case of failure in the torus window). The QO TL solution requires therefore an additional containment structure to satisfy safety requirements. A proposed solution that takes the advantages of the QO line and in principle solves the safety issues is an Evacuated QO (EQO), a MBTL enclosed in a vacuum vessel. The reference design is based on mirror confocal layout where the single unit is composed by a couple of mirrors forming a dogleg for TL bend and a straight path where the beams propagate alternatively crossing or parallel to each other. One pumping unit is foreseen for each unit. The characteristic length of the system L is defined as the distance between the two focusing mirrors. In order to validate this proposal solution a preliminary analysis on different aspects has been conducted. The theoretical absorbed power density on a mirror surface as a function of L has been calculated considering 8 Gaussian beams, each one of 2 MW (assuming a conservative mix of 50 % of either polarization), incident with a 45° angle on copper surface on vertices of a regular heptagon and one in the centre. The minimum beam envelope radius has been set $r = 1.5w + 90$ mm where w is the beam radius equal to $w = w_0*(1+(\lambda h/\pi w_0^2))^{1/2}$ and $w_0 = 20.43$ mm is the beam waist at the aperture of a waveguide of diameter 63.5 mm, λ the wavelength. The absorbed power density evaluated in different mirror points at 170/204 GHz is <0.3 MW/m^2 for L > 5/6.5 m and <0.2 MW/m^2 for L > 10/11 m) with minimum beam envelope radius of 0.2/0.24 m. The results are promising if compared with the same of ITER mirror mitre bend of ~2 MW/m^2. An overall estimation of theoretical losses for a generic EQO has been carried out starting from W7-X data [9]. Two different lengths have been selected, 100 m and 150 m, assuming L = 8 m. The transmission efficiency at 170 GHz is 88 % and 91 % for the respective lengths (the losses due to envelope wall were not considered), in according with the DEMO requirement of 90 % and in line with the same estimation done with ITER EWG components. Finally a preliminary cost analysis has been carried out evaluating three main contributions: mirrors, vacuum envelope and pumping system. The cost of components, based on recent quotations, has been related to distance *L*. The unit cost tends to be constant for $L > 6$ m and comparable with the EWG option.

4 System reliability and architecture

As a part of conceptual design phase the RAMI approach has been adopted and the impact of reliability requirement in a reactor like DEMO has been used as a guideline for designing the EC system architecture. The demanded reliability for an EC system should be as high as possible (assuming 99,9 % as a result of conservative consideration) and to reach this goal it is mandatory to introduce but minimize redundant elements to ensure consistency with the space required, the complexity of the control system and the system maintainability. The most critical condition is expected when full 50 MW of EC power is required. Since for a power plant the operating time should be the highest possible to ensure the cost effectiveness, the Mean Time Between Failures (MTBF) must be as high as possible, where a fault is the impossibility to deliver to the plasma the required 50 MW. In this study it was assumed as acceptable value 1000 pulses for the MTBF, here defined in terms of pulses with 2 h length each. With this assumptions the operating time of at least 2000 h between two faults corresponding to 3 months of DEMO operations. The basic configuration is a system composed by *l* number of simple EC lines (ECL) made up of 1 Power Supply Unit (PSU), 1 gyrotron, 1 TL and 1 launcher. For each element a reliability *R* is assumed: gyrotron $R_G = 98$ %, launcher $R_L = 99{,}9$ %, TL $R_{TL} = 99{,}9$ % and $R_O = 100$ % for other components (once defined the real values the calculation will be updated). The reliability of a single ECL is defined [10] as the product of the single reliabilities $R_{ECL} = R_G R_L R_{TL} R_O$ and the reliability of the system is:

$$R_{System} = \sum_{i=k}^{l} \binom{l}{i} R^i (1-R)^{l-i}$$

where *l* is the number of necessary lines to deliver 50 MW to the plasma and $k = 28$ is minimum *l* considering 10 % of TL losses to compensate. The minimum number of ECL to reach the target of MTBF > 1000 pulses is $l = 32$ with a MTBF = 1595 pulses. This solution is conceptually simple but leads to an EC system with a very large number of TLs: the volume occupied cannot be overlooked in the design phase and maintenance can be very expensive or impossible. To reduce this number an alternative solution consists in a few clusters in which *n* gyrotrons and *n* launchers are connected by single MBTL including *m* gyrotrons and *m* launchers spares for the reliability, 1 PSU and *n* Power Switches (with R_{PS} assumed 100 %). The key component of the cluster configuration is the TL. In this case the total reliability of each cluster can be expressed [10] as:

$$R_{Cluster} = R_O \left(\sum_{i=n-m}^{n} \binom{n}{i} R_G^i (1-R_G)^{n-i} \right) * R_{TL} \left(\sum_{i=n-m}^{n} \binom{n}{i} R_L^i (1-R_L)^{n-i} \right)$$

Two options are viable: $m = 0$ cluster without backup components where the total reliability is improved increasing the number of clusters (as simple ECL configuration) and $m > 0$ cluster with backup component where the total reliability is enhanced by increasing the reliability of each cluster.

Table 3. Cluster configuration results of ECS reliability with 1 backup item.

Number of gyrotrons/ launchers per cluster [*n*+*m*]	Number of clusters [*j*] to deliver 50MW	R_{system} [%]	MTBF [pulses]	Total number of gyrotrons/ launchers
1+1	28+1	99,9601	2507	58
2+1	14+1	99,9896	9606	45
3+1	10+1	99,9945	18291	44
4+1	7+1	99,9972	35852	**40**
5+1	6+1	99,9979	47777	42
6+1	5+1	99,9985	66830	42
7+1	**4+1**	**99,9987**	**79870**	**40**
8+1	**4+1**	99,999	100198	45
9+1	**4+1**	99,999	100200	50

Table 3 shows the results for cluster with the use of one backup item (*m*=1). Two solutions with the minimum total number of 40 gyrotrons and launchers are possible: the first with $n+m = 5$ (4+1) and $j = 8$ (7+1) MBTLs, and the second with $n+m = 8$ (7+1) and $j = 5$ (4+1) MBTLs. The solution with $j = 8$ (7+1) is preferable because it results a good trade-off between a lower number of clusters (only 5) and a higher reliability and MTBF. The best solution is composed by only 5 MBTLs, each of them connected with 8 gyrotrons and 8 launchers (1 for both spare). Analogous result has been obtained with the analysis of $m = 0$ case. The cluster solution with $m = 1$ also reduces the number of subsystems, saves space with respect to the single ECL and the high MTBF assures safe margins when the real reliability of other single component (R_O) will be defined.

5 Conclusions

The DEMO EC system will be responsible for several key physical tasks but with much more limited flexibility compared to present experimental devices. Different options for gyrotrons, TLs and launchers are under evaluation taking into account the RAMI target requirement for a nuclear power plant. Multi-frequency

2MW gyrotron with high efficiency and reliability is the present source assumed. The study of RSA was approached whereas other simplified antenna types (e.g. TWG) will be considered at later stage. A preliminary calculation of blanket apertures has been conducted and the relative low impact on TBR evaluated and confirmed. The new concept of EQO MBTL can be a viable option for a DEMO reactor and adopted for system in cluster units as shown by analysis reaching the high required reliability. The details of EC system have still to be fixed but a design method in a reactor compatible way is progressing.

This work has been carried out within the framework of the EUROfusion Consortium and has received funding from the Euratom research and training programme 2014-2018 under grant agreement No 633053. The views and opinions expressed herein do not necessarily reflect those of the European Commission.

References

1. F. Romanelli et al., EFDA fusion roadmap
2. S. Garavaglia et al., AIP Conference Proceedings **1689**, 090009 (2015)
3. M. Kovari et al., Fusion Eng. and Des. **89**, 12, 3054-3069 (2014)
4. J.F. Artaud et al., in 32nd EPS Conf. on Control. Fusion and Plasma Phys., ECA Vol. 29C, P1.035 (2005)
5. M. Romanelli et al., Plasma and fusion research **9**, 2, 3403023 (2014)
6. G. Denisov et al., This Proceedings
7. J. Jelonnek et al., This Proceedings
8. E. Poli et al., Computer Physics Communications, **136**, 1–2, 90–104 (2001)
9. V. Erckmann et al., Fusion Sci. Technol. **52**, 291-312 (2006)
10. A. Birolini, Reliability Engineering: Theory and Practice 7th Edition Springer

Series of powerful CW gyrotrons in the range 105 – 140 GHz

Gregory *Denisov*[2], Igor *Kazansky*[1], Vladimir *Malygin*[2], Elena *Soluyanova*[1], Evgeny *Tai*[1*], Aleksey *Chirkov*[2]

[1]Gycom Ltd, Nizhny Novgorod, Russia
[2]Institute of Applied Physics, Nizhny Novgorod, Russia

Abstract. GYCOM in collaboration with Institute of Applied Physics (IAP) designed, fabricated and tested two different gyrotrons with similar to ITER parameters at 105 – 140 GHz frequency range. Set of gyrotron system beside the tube includes matching optic unit (MOU) and elements of evacuated transmission lines. One gyrotron system was installed and operates at plasma machines EAST / ASIPP / China, another system is delivered to KSTAR / NFRI / Korea.

1 Introduction

Presented gyrotrons are similar to ITER gyrotrons in points of RF power up to 1 MW at the output window, duration of microwave pulses up to 1000 s, efficiency about 50 %. The gyrotron systems include matching optic unit (MOU) for HE11 waveguide mode forming, calorimetric load for full regime of operation, special pumping tee, RF two-directional switch, corrugated waveguides, miter bends and miter bend with bi-directional coupler.

First of the mentioned gyrotrons was used at EAST machine for plasma heating experiments during 2015 - 2016 years. It has 140 GHz frequency, 1 MW power and maximum 1000 s longevity of CW operation. Second gyrotron is a two-frequency 105 / 140 GHz tube designed for plasma machine KSTAR. Output power of this tube is 950 kW at frequency 140 GHz and 800 kW at 105 GHz, maximum pulse duration is 300 s. The parameters of two-frequency tube were demonstrated already at GYCOM's factory tests. Now equipment is under acceptance test procedure at the customer site. Set of delivery besides the tube includes two-frequency MOU with removable mirrors and calorimetric load. Both gyrotron systems use for operation standard liquids-free magnet JMTD-7T160 produced by company JASTEC, Japan. The series of gyrotrons will be continuing by new contracts with ASIPP and NFRI.

2 Features of gyrotron systems design

Main feature of described gyrotrons consists in scheme of CPD insulation and is shown at the Fig.1. Classic ITER-like gyrotron has located above cryomagnet CPD ceramic, its body inside cryomagnet has positive relatively ground potential and correspondently is electrically insulated from cryomagnet warm bore. Described in this paper gyrotrons are based on located under cryomagnet CPD ceramic and have grounded tube body without HV insulation between gyrotron and magnet. "Lower" CPD insulator design differs by smaller insulator diameter (140 mm against 240 mm diameter in "upper" ceramic) and correspondently more simple system for ceramic cooling. Disadvantages of "lower" CPD insulator are smaller maximum recuperating voltages, correspondently smaller maximum efficiency (theoretically difference of efficiency is about 3%) and necessity of vertical transportation because of anode-cavity unit of gyrotron has long cantilever design.

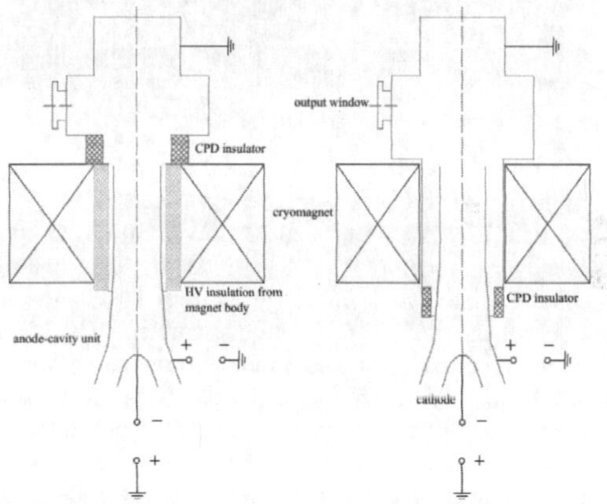

a b

Fig. 1. Scheme of gyrotrons supplement and insulation, a – "upper" CPD insulator, b – "lower" CPD insulator.

Other features of presented gyrotron tubes design are: stainless steel double wall body cooled by water between walls, cooled by water anti-arcs screen in cathode area, high temperature LaB_6 emitter at cathode with middle diameter of 86 mm, relief BN window with 100 or 123 mm diameter, collector with combination of DC and AC sweeping coils. All presented gyrotrons are equipped by brazed by GYCOM diamond windows with aperture from 70 to 88 mm. Diamond disks for windows are produced by E-6 or by IAP and brazed by surface-type connection with direct diamond cooling or edge-type

* Evgeny Tai: tai@gycom-nn.ru

connection with diamond cooling via copper cuff. Cavity modes are $TE_{22,8}$ for 140 GHz operation (for both types of tubes) and $TE_{17,6}$ for 105 GHz (for two-frequency tube). Additional feature of two-frequency gyrotron system is utilization of removable mirrors in MOU design for minimization of time for frequency changing and convenience of application.

3 Experimental results

Gyrotron systems were tested at GYCOM factory test bench. The scheme of the test is shown at Fig.2. Transmission line was pumped out through two ports. First port is located at MOU, second port is located at pumping tee. The pressure in transmission pump without RF power is less than 10^{-5} mbar. Vacuum two-directional switch directs the wave beam at short pulse calorimetric load with barrier window or at the full regime CW load.

Fig. 2. Gyrotron system assembled for the tests.

During 140 GHz / 1 MW / 1000 s gyrotron factory tests following regimes were demonstrated: 0.5 MW / 1000 s, 0.75 MW / 500 s, 0.92 MW / 300 s. Output power were measured at CW load by calorimetric method. Cathode voltage was varied from 44.5 to 46 kV, anode voltage – 22-23 kV, cathode current – 30-35 A. Traces of operation at maximum power are shown at Fig.3. It should be noted that experiments with gyrotrons provided by high temperature cathode were fulfilled without feed back system in the filament circuits. It causes some decreasing of cathode current after pulse start. The time of beam current stabilization is about 50 s and after that all operating parameters became stable. The calorimetric signals of powerful operation regime are presented at Fig.4.

Two-frequency 140 / 105 GHz gyrotron design differs from 140 GHz tube by cathode position. The gyrotron operation on both frequencies is accompanied by magnetic field correction in cathode area by using of special coil. Operation regimes at 140 GHz in both gyrotrons are very similar. Regime of operation at maximum power of 780 kW in the load at frequency 105 GHz is shown at Fig.5. Change of the gyrotron system frequency supposes following actions: change main magnetic field direction and magnetic field direction in AC section of collector coil, inject the atmospheric pressure in MOU and transmission line, change mirrors in MOU, pump out the MOU and transmission line, adjust regime of operation. This procedure takes in total about 12 hours.

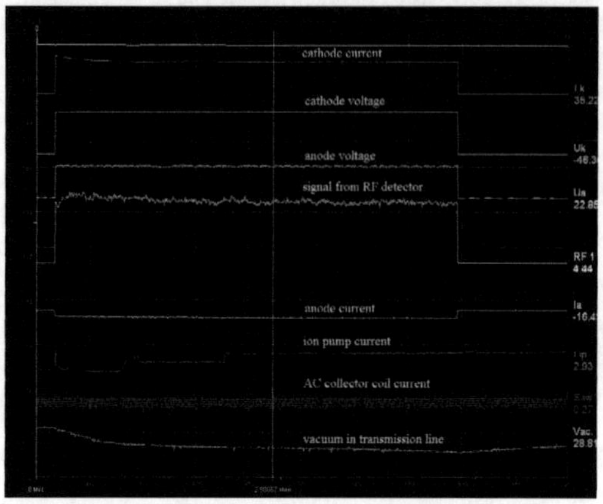

Fig. 3. Traces of 140 GHz operation at 0.92 MW / 300 s.

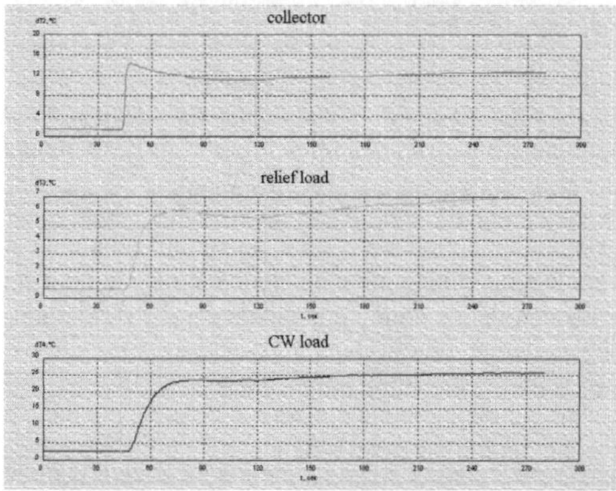

Fig. 4. Calorimetric signals of 140 GHz operation at regime 0.92 MW / 300 s.

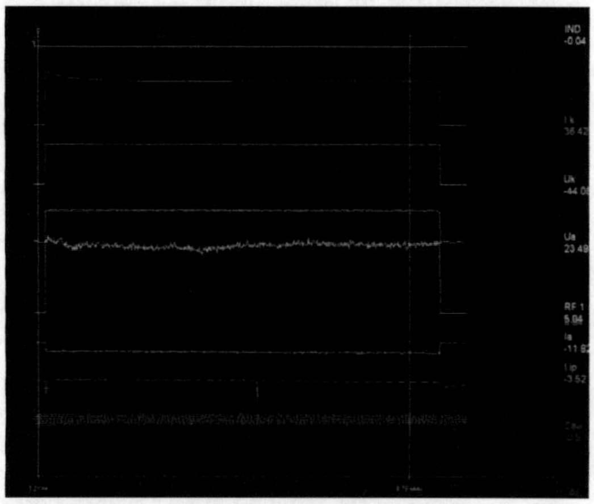

Fig. 5. Two-frequency gyrotron operation at 105 GHz. Cathode current Ik=36.4 A, cathode voltage Uk=44 kV (negative to ground), anode voltage Ua=23.5 kV (positive to ground), power P=780 kW (in the load).

Remote-Steering Antennas for 140 GHz Electron Cyclotron Heating of the Stellarator W7-X

C. Lechte[1,a], W. Kasparek[1], B. Plaum[1], A. Zeitler[1], V. Erckmann[2], H. Laqua[2], N. Schneider[2], M. Weissgerber[2], A. Bechtold[3], M. Busch[4], and B. Szepaniak[4]

[1] *Institute of Interfacial Process Engineering and Plasma Technology IGVP, University of Stuttgart, Germany*
[2] *Max-Planck-Institut fuer Plasmaphysik, Boltzmannstr. 2, 85748 Garching, Germany*
[3] *NTG Neue Technologie GmbH & Co KG, 63571 Gelnhausen, Germany*
[4] *Galvano-T electroplating-electroforming GmbH, 51570 Windeck-Rosbach, Germany*

Abstract. For electron cyclotron resonance heating of the stellarator W7-X at IPP Greifswald, a 140 GHz/10 MW cw millimeter wave system has been built. Two out of 12 launchers will employ a remote-steering design. This paper describes the overall design of the two launchers, and design issues like input coupling structures, manufacturing of corrugated waveguides, optimization of the steering range, integration of vacuum windows, mitrebends and vacuum valves into the launchers, as well as low power tests of the finished waveguides.

1 Introduction

The ECRH system at the stellarator Wendelstein 7-X (W7-X) has 10 MW of continuous power at 140 GHz [1]. The beams from the 10 gyrotrons are transmitted via 2 quasi-optical multi-beam transmission lines. In addition to the conventional equatorial launchers with front steering mirrors, launchers in the N ports allow quasi-high field side injection into the plasma. However, since the N ports are much smaller, it was not feasible to use front steering mirrors.

The remote steering principle uses corrugated square waveguides which image a tilted beam from the input to the output side. At the output, no movable parts are necessary and the waveguide can end flush with the vessel wall. This makes the remote steering solution very compact and suitable for use in a fusion power plant.

In W7-X, 2 beams can be rerouted from the front steering launchers to the remote steering launchers (RSL1 and RSL5, see Fig. 1.) The RSLs are water cooled for cw operation and they incorporate vacuum elements (diamond window and plate valve) as well as a mitre bend.

The design and manufacture of the RSLs were done within the framework of the FORMIK[3] project with financial support from the German Federal Ministry for Education and Research. The main components were fabricated by the industrial partners NTG and Galvano-T.

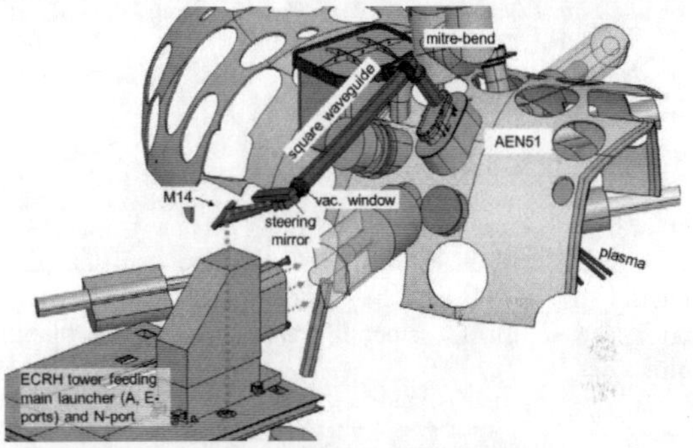

Figure 1. Concept of the remote steering launcher RSL5 for W7-X

2 The remote steering principle

The remote steering method using corrugated square waveguides employs the Talbot effect [2] to image the input pattern at the output side. This effect takes place if the length L of the waveguide with width a is as in Eq. (1). If the input pattern represents a Gaussian beam with tilt or steering angle Φ, then the output pattern represents the same beam but with steering angle $-\Phi$. This is shown in Fig. 2. The beam pattern is made up of several waveguide modes which propagate with different phase speeds. At the end of the waveguide, the phase differences are an integral number of π again and the pattern in recreated.

Since the imaging effect depends on a paraxial approximation, it breaks down for large (>12°) angles (Fig. 2, bottom plot.) This manifests as an incorrect

[a] Corresponding author: carsten.lechte@igvp.uni-stuttgart.de

length of the waveguide. A formula that corrects for the steering angle Φ is given in Eq. (2).

$$L = 4\,a^2 / \lambda \qquad (1)$$

$$L_{corr} = L / (1 + 0.5 \sin^2 \Phi) \qquad (2)$$

As a consequence, for larger steering angles, the RSL needs to be slightly shorter [3]. For a steering angle of 22.4°, the waveguide is long enough for the beam to complete another internal reflection and produce an output beam that goes into the opposite direction (revival).

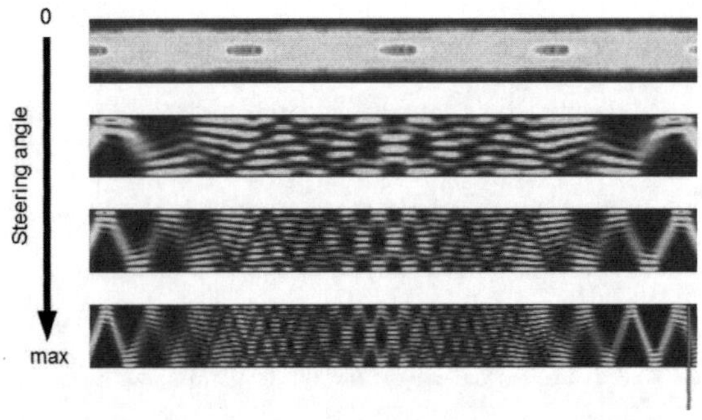

Figure 2. Field pattern in a corrugated rectangular waveguide for tilted input beams. For large angles, the beam shifts so far off-centre, that it hits the waveguide wall.

3 Design for W7-X

In order to fit into the W7-X ECRH system, the RSL waveguides needed to incorporate a mitrebend near the middle of its length. Its corrugation was carefully optimised to form a smooth transition between the 2 directions.

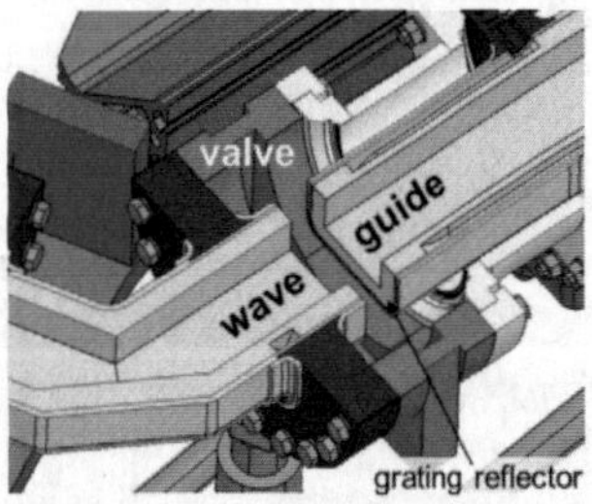

Figure 3. Incorporation of vacuum valve at the mitrebend.

The inside would be under vessel vacuum, with a diamond vacuum window at the input side. An additional vacuum valve was necessary in case of window failure. Since this leaves a 20 mm gap in the waveguide, it was located at half length, where the field at the waveguide walls is minimal and therefore stray radiation is minimised. A grating reflector keeps stray radiation away from the heat sensitive vacuum valve O rings.

Because of the large Brillouin angle of the beam inside the waveguide (see Fig. 2), ohmic losses are expected to be in the 1-2 % range when using copper material. Therefore, water cooling is mandatory to enable cw operation.

The width of the waveguides is approximately 50 mm and the length approximately 5 m.

4 Optimisation of the corrugation profile

To design a corrugation profile that meets high power cw specifications, extensive fullwave simulations were made [4]. For the hybrid modes in the waveguide to propagate correctly, a corrugation period of 0.8 mm and an effective depth of λ/4 is necessary. The profile must have a rounded shape to avoid arcing, but still must have a 180° phase difference between the TE and TM modes. Finally, to minimise ohmic losses, the grooves should be wider than the ridges. The waveguide was manufactured by electroforming, i.e. the negative form was machined in aluminium and the waveguide made from electroformed copper.

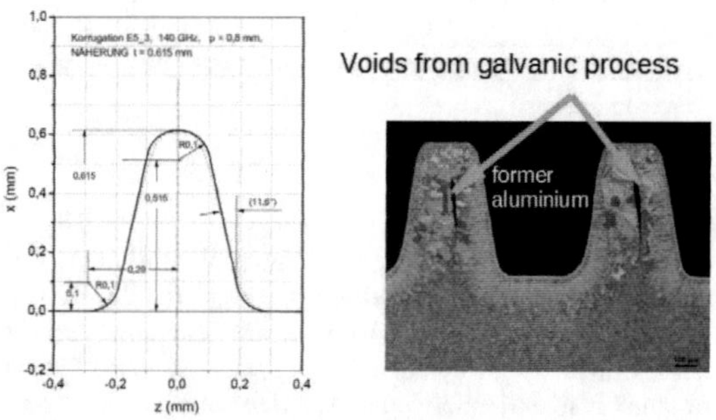

Figure 4. Corrugation profile (left) and cut of a test piece (right) after electroforming.

The chosen corrugation profile is shown in Fig. 4. Because of the manufacturing process and the cooling (heat conduction) requirements, the ridges could not be made very thin. Even so, voids have formed in the material of the ridges because of the complicated growth pattern of the electroformed copper inside the grooves of the aluminium mandrel. The voids were found to be unproblematic for the functioning to the waveguide.

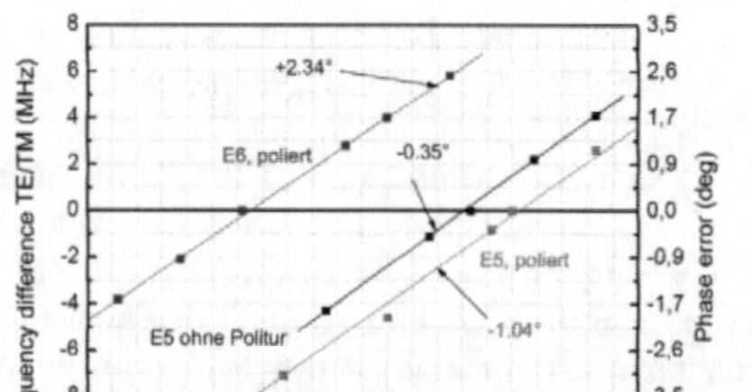

Figure 5. Measurement of the phase error between the 2 polarisations at various frequencies.

5 Optimisation of the steering range

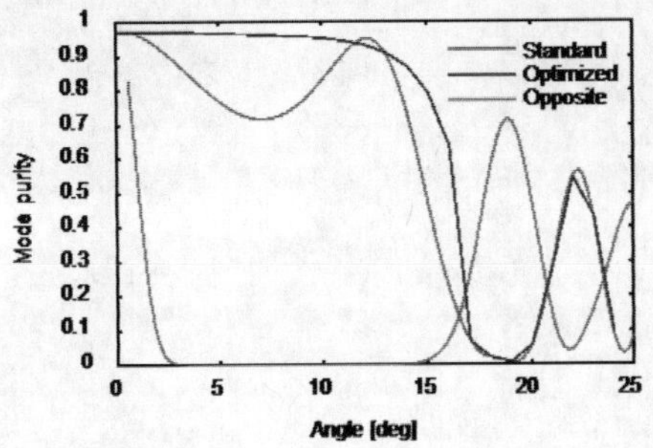

Figure 6. Mode purity of the output Gaussian beam for different steering angles with an optimum waveguide length for 12.7°.

The non-optimised usable steering angle range of 12° was unsatisfactory for the W7-X system. A variable mechanical length change according to the current steering angle is not feasible for a water-cooled, corrugated waveguide. Therefore, a virtual length adjustment was designed by way of an optimised input coupling. The increasing lateral displacement of the beam at large steering angles was counteracted by a lateral displacement at the input side, which is equivalent to shifting the intercept of the input beam with the waveguide axis into the waveguide, effectively shortening it to the ideal length given in Eq. (2).

In addition, the length of the waveguide was set to the ideal length for a steering angle of 12.7°. The resulting deterioration of the beam quality for lower angles (see red curve in Fig. 6) was again compensated by the optimised input coupling, which shifts the beam in the other direction, effectively lengthening the waveguide. The result can be seen in Fig. 6, blue curve, which shows a usable beam from 0° up to 16° steering angle. Finally, the revival of the beam in the opposite direction is shown in green.

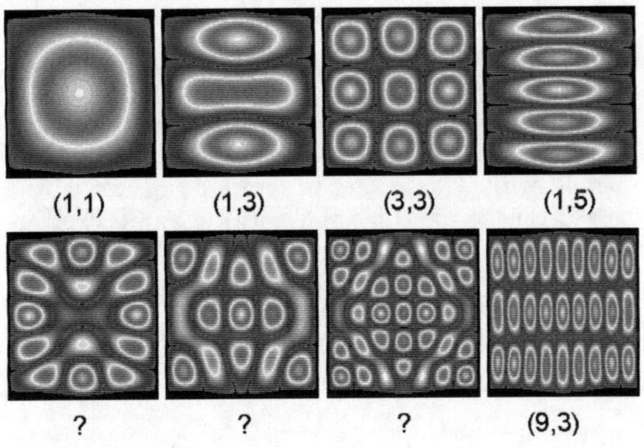

Figure 7. Perpendicular mode patterns in the deformed waveguide cross section for A=2 mm and q = 4.

A different approach to the improvement of the steering range was the modification of the dispersion relation inside the waveguide itself. This can be achieved by deformation of the cross section of the waveguide. The deformations were of the shape A $\cos(x)^q$ where A is the amplitude (up to a few percent of the waveguide width) and q is an integer between 2 and 5. The modes and propagation constants of the deformed waveguide were calculated with the IPF-FD3D [5] code.

Some resulting modified mode patterns are shown in Fig. 7. While some modes have clear correspondences to the undeformed modes, some patterns are not. Together with the modified propagation constants, the imaging properties of the deformed waveguide could be simulated with the PROFUSION code.

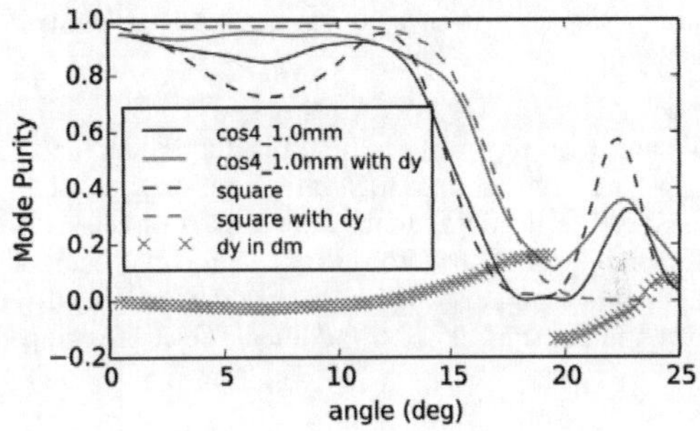

Figure 8. Mode purities of the deformed vs. the undeformed waveguides. dy is the lateral displacement from the optimised input coupling.

As can be seen in Fig. 8, the best deformation found gives only a moderate improvement of the beam quality, even when combined with the optimised input coupling. Indeed, the best result comes from using the undeformed square waveguide with the optimised input coupling. It was not possible to translate the good improvement found in [6] for an ITER RSL to 140 GHz. While this result is disappointing, the investigations are still being pursued with other classes of deformations.

6 Fabrication

The 2 RSLs were fabricated at the industrial partners. In order to achieve vacuum tightness and high quality corrugations, a galvano-plastic (electroforming) technique was used. For each RSL, the 2 straight parts and the mitrebend were manufactured as one piece, respectively.

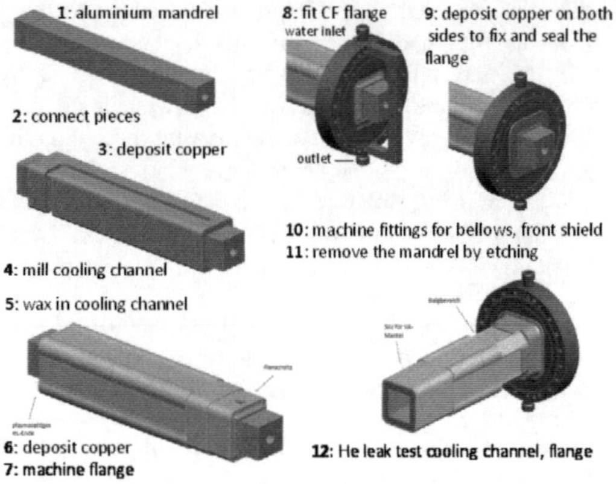

Figure 9. Electroforming fabrication process for the corrugated square waveguide (length is understated for illustration purposes.)

The negative form was machined in aluminium, and the copper material subsequently galvanised on top (see Fig. 9). Specialised milling tools were built to precisely and efficiently mill the corrugation grooves into the aluminium mandrel. Cooling channels were milled into the aluminium and closed by additional electroforming.

7 Results of low power tests

The finished RSL1 and RSL5 were transported to IGVP for low-power testing and beam characterisation. Fig. 10 shows the setup of RSL1 with the input mirror and diamond window. For each waveguide, the full setup with 2 straight pieces, the mitrebend, vacuum valve, diamond window, and coupling optics was set up. A vector network analyser was used as the source, with the receiving antenna mounted on a 2D movable platform at the output side.

The far field measurements (at a distance of 1 m) are shown in Fig. 11 for various steering angles. The output beam is clearly defined, and the side lobes are moderate. A similar result was obtained for RSL5.

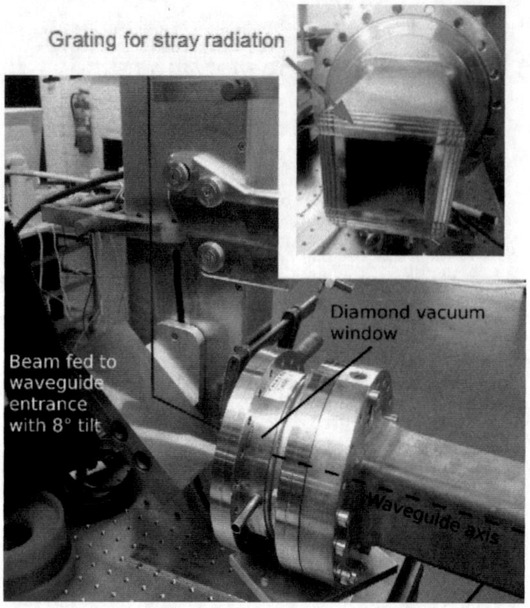

Figure 10. Input side of the finished RSL1 with optimised input coupling optics. The inset shows the other end of the straight waveguide piece.

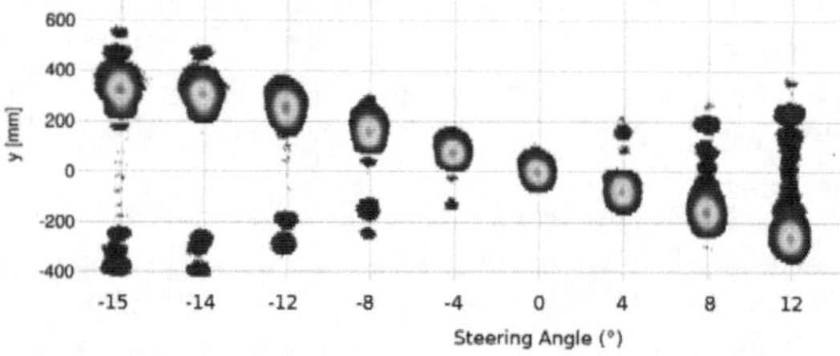

Figure 11. Far field measurements for RSL1. The steering effect is clearly seen and the side lobes are moderate (linear scale).

The output beam quality was assessed using near field measurements directly at the mouth of the waveguide. A fine spatial sampling was used to capture both the amplitude and the phase of the beam. A Gaussian beam was fitted to the data. The fraction of total power in this Gaussian beam was taken as the efficiency of the beam.

At 0°, a Gaussian with a waist size of 15.6 and 14.5 mm for E and H plane was found, respectively. The phase front was flat. For the optimised angle 12.7°, a phase front tilt of 12.6° was found, in very good agreement with the expected beam. At the optimised angle, it is expected that the beam is perfectly centred on the waveguide axis (no lateral shift necessary). However, a 5 mm shift was found. The shift only disappeared at 14.5°, which would hint that the waveguide was effectively shorter and its optimal steering angle therefore larger.

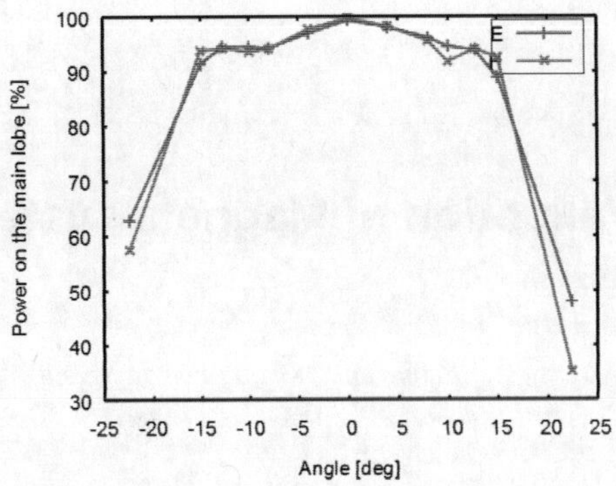

Figure 12. Beam quality measurements in the near field, E and H plane.

Nevertheless, the beam quality (shown in Fig. 12) was still very good for steering angles between -15° and 15°. Only the revival beams (expected at 22.4°) were unusable. Due to the effective shortness of the waveguide, their angles, too, would be shifted to larger values.

The explanation for this shortening of the waveguide can be found be examining the effective waveguide width a: For a square corrugation profile, the real and the effective groove depth are both $\lambda/4$, but for the rounded profile, the actual profile depth is slightly larger than that to have the same effect. If the wave then penetrates slightly into the grooves, the effective waveguide width increases, which means that the nominal length L also increases (see Eq. (2)). Since this was not taken into account, the waveguide is a few cm too short, which explains the observed behaviour.

The design of the optimised input coupling has been updated to take the new optimal angle into account.

8 Conclusions

The BMBF project FORMIK[3] was brought to a successful conclusion. The performance of the 2 RSLs was characterised with low power measurements and found to be good with moderate side lobes. They were delivered to W7-X, where they are awaiting installation and high power tests.

The design and manufacturing experience gained will be very valuable for DEMO where remote steering has to be considered because of the harsh reactor environment and the dearth of large ports.

Acknowledgements

This work is supported by the German Ministry of Education and Research (BMBF) by grant 03FUS0017A.

References

1. V. Erckmann et al., Electron cyclotron heating for W7-X: Physics and Technology. Fusion Science and Technology **52**, No 2 (2007) 291-312.
2. W.H.F. Talbot, Facts relating to optical sciences. No. IV. Philos. Mag. 9 (Third series), **401** (1836).
3. K. Ohkubo et al., "Extension of steering angle in a square corrugated waveguide antenna". Fusion Eng. Des. **65** (2003), 657 – 672.
4. B. Plaum et al. "Numerical calculation of reflection characteristics of grooved surfaces". J. Infrared Milli. Terahz. Waves **32** (2011), 482–495.
5. C. Lechte "Investigation of the Scattering Efficiency in Doppler Reflectometry by Two-Dimensional Full-Wave Simulations" IEEE Trans. Plas. Sci. **37** No 6, 2009, 1099-1104
6. G. G. Denisov et al: "Efficiency enhancement of components based on Talbot effect". Int. J. Infrared Milli. Waves 28 (2007), 923–935

Design, Development & Functional Validation of Magnets system in support of 42 GHz Gyrotron in India

S. Pradhan, P. Raj, U. Prasad, M. Ghate, Y. Khristi, A. Panchal, D. Bhavsar, M. Banudha, S. Kedia, A.N. Sharma, D. Kanabar, B. Parghi

Institute for Plasma Research, Bhat, Gandhinagar, Gujarat-382428

Abstract. A multi institutional initiative is underway towards the development of 42 GHz, 200 kW gyrotron system in India under the frame work of Department of Science and Technology, Government of India. Indigenous realization comprising of design, fabrication, prototypes and functional validations of an appropriate Magnet System is one of the primary technological objective of these initiatives. The 42 GHz gyrotron magnet system comprises of a warm gun magnet, a NbTi/Cu based high homogenous superconducting cavity magnet and three warm collector magnets. The superconducting cavity magnet has been housed inside a low loss cryostat. The magnet system has been designed in accordance with gyrotron physics and engineering considerations respecting highly homogenous spatial field profile as well as maintaining steep gradient as per the compression and velocity ratios between the emission and resonator regions. The designed magnet system further ensures the co-linearity of the magnetic axis with that of the beam axis with custom winding techniques apart from a smooth collection of beam with the collector magnet profiles. The designed magnets have been wound after several R & D validations. The superconducting magnet has been housed inside a low loss designed cryostat with in-built radial and axial alignment flexibilities to certain extent. The cryostat further houses liquid helium port, liquid nitrogen ports, current communication ports, ports for monitoring helium level and other instrumentations apart from over-pressure safety intensive burst disks etc. The entire magnet system comprising of warm and superconducting magnets has been installed and integrated in the Gyrotron test set-up. The magnet system has been aligned in both warm and when the superconducting cavity magnet is cold. The integrated geometric axes have been experimentally ensured as well as the field profiles have been measured with the magnets being charged. Under experimental conditions, all magnets including the superconducting magnet have been charged to their nominal values with appropriate protection measures against the quench. This is the first time in India that a gyrotron specific magnet system with superconducting magnet has been realized.

1 Introduction

Gyrotron is a vacuum tube capable of generating high power of few millimeter-wavelength beams by electron bunching. These high power electron beams can be used to raise plasma temperature. The Electron Cyclotron Resonance Heating (ECRH) is one of the applications of gyrotron. It is basically a fast-wave microwave tube based on the principle of interaction between the gyrating beam and the transverse electric (TE) mode [1-2]. The development of gyrotron devices are focused on to produce high-power, continuous-wave, or long-pulse as a power source for Electron Cyclotron Heating (ECH). If the bunch of electron waves has correct frequency and polarization, then their energy can be transferred to the charged particles in plasma, which in turn collide with other plasma particles, thus increasing the temperature of the bulk plasma. Design, development of required technologies and its validation towards indigenous manufacturing of long pulse gyrotron is an important milestone to develop fusion grade gyrotron system in the fusion roadmap of India [1-3]. With these motivations, a multi-institutional project under department of science and technology (DST), India has been initiated to develop country's first indigenous long pulse gyrotron. The major specification of 42 GHz, 200 kW Long Pulse gyrotron is given in Table.1. Magnet division of Institute for Plasma Research (IPR) is primarily responsible for the design and manufacturing of Gyrotron Magnet System (GMS). The design and optimization of GMS has been completed and reported. [1-2]. The Fabrication steps, experimental test setup and testing methodology for individual magnets in GMS and its assembly as per requirement is discussed in this paper. The magnetic field homogeneity for individual magnet as well as assembled GMS is mapped in radial as well as axial direction. The details of fabrication, testing, assembly, optimization and alignment and long pulse operation of GMS have been discussed in this paper.

* Corresponding author: pradhan@ipr.res.in

 DOI: 10.1051/epjconf/201714704005

Table 1. Major specification of gyrotron

Particulars	**Specifications**
Frequency	42±0.05 GHz
Output Power	200 kW
Pulse Duration	Long Pulse (Max. 3000 ms)
Efficiency	35%
Wall Loading	< 2.0 kW/cm^2
Beam Voltage	65 kV
Beam Current	10 A
Gun Type	Magnetron Injection Gun
Interaction Circuit	Weakly Tapered Cavity

2 Design of Gyrotron Magnet System

The magnetic field required for the 42-GHz gyrotron at the interaction region is calculated to be around 1.64 ± 0.04 T for exciting $TE_{0,3}$ mode during interaction. The variation in the field profile at the interaction region is ±20 mm and less than ±0.01 T (±1%) in axial direction. A very precise and homogeneous axial magnetic field is required along the path of the electron beam. For the 42 GHz gyrotron, a maximum tolerance of ±5% is allowed in the field profile along axial and radial direction except near interaction region. In gyrotron, three regions (beam trigger, beam interaction and beam collection region) are important to generate high energy electron beam. In order to meet such conditions, gyrotron magnet system consists of three kinds of magnets (Gun Magnet, Cavity Magnet and Collector Magnet) as shown in Fig.1. Gun magnet is used to optimize the quality of an electron beam by simultaneously maximizing the electron orbital-to-axial velocity ratio without reflected electrons and minimizing the velocity spread [4-5]. Cavity magnet provides very high homogeneous field which is required for the beam–wave interaction of microwave. Three collector magnets are used to optimize the beam power deposition at collector surface. Collector magnets are used to optimize the beam power deposition at collector surface. To reduce beam power density deposition at the collector, these coils are fed with alternating current and, therefore also called Saddle or sweeping coil. The design and fabrication of magnets have been done under consideration of field profile homogeneity along axial (z-direction) as well as in radial direction (r-direction). The superconducting magnets for gyrotron application employ high current carrying and high field multi-filamentary technical superconductors, such as multi-filamentary NbTi strand that has been procured for the manufacturing of low T_C solenoid cavity magnet mainly for the high field interaction region of the gyrotron [1-2]. In few cases, intermediate superconductors such as MgB_2 and high T_C Bismuth Strontium Calcium Copper Oxide (BSCCO) and Yttrium Barium Copper Oxide (YBCO) have also been attempted [6-7].

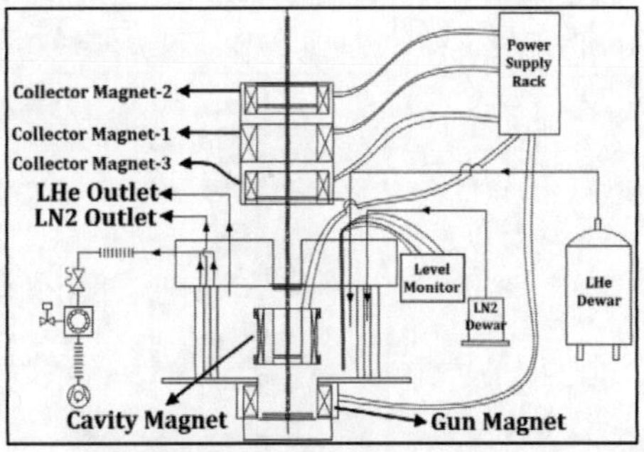

Fig.1. Schematic of Gyrotron System

Since the requirement of the field profile is very specific, therefore required magnetic field profile along z-direction is optimized by taking into account the magnets locations and their engineering and fabrication limitations [8-9]. Since the development of the gyrotron device is taking place for the first time in the India, it is necessary to have flexibility in all the components towards some adjustments in case of any unforeseen requirements. The magnet system is no exception. Thus, a somewhat adjustable/variable magnetic field profile is required at the time of operation for slightly tuning the magnetic peaks to get optimum output. The variation in any parameter except the magnet transport current is not feasible, once the magnets are fabricated and mounted inside the housing cryostat and oil tank. However, the gun and collectors coils currents may also have to be adjusted to maintain the beam quality and the power density deposition rate. Hence, the variation in the entire field profile is slightly nonlinear for the desired range of ±5%, which is acceptable from gyrotron physics considerations.

3 Manufacturing of Gyrotron Magnet System (GMS)

The fabrication of superconducting magnets ensures not only an appropriate gradient of the magnetic field in the accelerator region but also high homogeneity in the resonator (interaction) region for optimal electron beam interaction. The cavity magnet fabrication also ensures the protection of superconducting magnets in case of irreversible normal zone evolution during operation known as quench. Due to quench, the rate of rise of temperature in the magnet winding pack following the maximum temperature known as hot-spot temperature are also critical criteria. Following the quench, a high voltage is also generated across the terminals of the magnet, and accordingly, the insulation (turn-to-turn, layer-to-layer, and ground) has been decided. An enameled NbTi strand has been used to fabricate cavity magnet. The cavity magnet has been fabricated as a combination of three solenoid magnets (a long solenoid magnet and two small magnets in Helmholtz-type configuration) at interaction region of gyro tube. This configuration has helped to get the required homogeneity of magnetic field at interaction region. The quench event of the magnet is promptly diagnosed with a smart quench detection logic and front-end electronics. The stored energy in the magnet is allowed to be distributed inside the winding pack with respect to the hot spot temperature criteria known as passive protection. All resistive magnets (Gun Magnet or MIG Magnet and Collector Magnet) have been fabricated with enameled Cu strands. Fiber glass tapes has been used as the inter layer, inter turn and ground insulation in all the magnets. The magnet winding pack has been vacuum pressure impregnated with epoxy resin in a special vacuum pressure impregnation facility. Glass fiber (G-10) pieces are used as a support material within the magnet winding where ever required. The approximate amount of insulation will be around 20% of the total volume of the magnet.

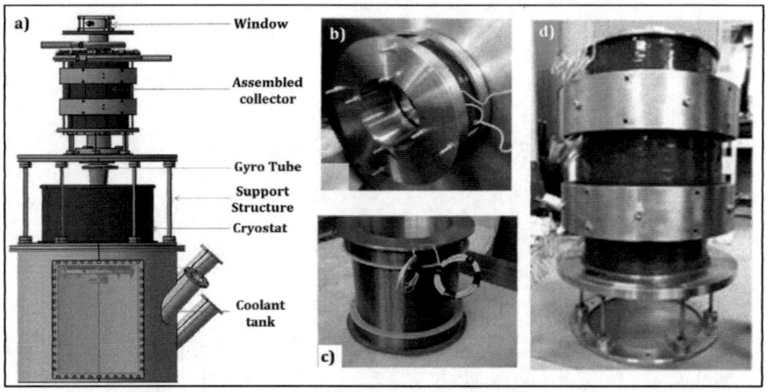

Fig. 2.a) Gyrotron magnet system b) Gun magnet c) Cavity magnet as wound d) Assembled collector magnets

3.1 Gun Magnet

Gun magnet has been fabricated with the 2.1 mm diameter copper strand. The winding pack of gun magnet has approximately 1120 turns. The detail specification and gun magnet is shown in Fig. 2b and Table.2. Gun magnet can produce magnetic field of maximum 0.03 T at current of 6.67 A.

Table 2. Technical specifications for gun magnet or MIG magnet

Description	Parameters
Winding Pack inner radius	135 mm
Winding height	120 mm
Number of layers	20
Number of total turns	~1120
Conductor material	Enamelled Cu
Wire strands diameter	2.1 mm
Insulation voltage capacity of strands	> 500 V
Inter turn voltage capacity	200 V
Ground insulation voltage capacity	1 kV

3.2 Collector Magnet

Collector magnet is manufactured using enameled copper strands of diameter 2.1 mm in layered configuration with required number of 2100 and 1000 turns respectively. For a 42 GHz gyrotron, field required at the collector magnet is ~1000 G. Collector magnet assembly consists of three individual magnets of two different types as shown in Fig.2d. The detail specification for collector magnets is given in Table 3.

Table 3. Technical specifications for collector magnet

Description	Collector Type 1	Collector Type 2
Winding pack inner radius	135 mm	135 mm
Winding height	150 mm	110 mm
Number of layers	~ 30	~ 20
Number of total turns	~ 2100	~ 1000
Conductor material	Enamelled Cu	Enamelled Cu
Wire strands diameter	2.1 mm	2.1 mm

Voltage capacity of strands	>500 V	>500 V
Inter turn voltage capacity	> 200 V	> 200 V
Ground insulation voltage	1 kV	1 kV

Table 4. Technical specifications for cavity magnet

Description	Parameters
Winding pack inner radius	105 mm
Winding height	130 mm
Number of layers (Solenoid)	~ 24
Number of total turns (Solenoid)	~6000
Number of layers (Each Notch)	15
Number of total turns (Each Notch)	~630
Conductor material	Multi-filamentary NbTi
Strand diameter	0.75 mm
Insulation voltage capacity of strands	>500 V

3.3 Cryo Magnet/Cavity Magnet and Low Loss Cryostat

The Cavity region magnet provides a very high homogenous magnetic field which is required for the electron beam interaction at bore of microwave tube. Cavity magnet consists of superconducting solenoid magnet manufactured using NbTi strands housed inside a low loss cryostat. The specification for cavity magnet is given in Table. 4. Fig. 2 C and 3 show superconducting magnet along with its positioning inside the cryostat. It has been designed to reduce loss of liquid helium. It is a multi-chamber cryostat designed to reduce heat load on superconducting magnet. The cryostat is provided with liquid nitrogen (LN_2), liquid helium (LHe) filling port, current leads, vacuum ports and recovery lines for gaseous helium. [10]

This cryostat has multiple chambers for vacuum, LHe chamber and LN_2 thermal shield. In Fig.3 it is shown that there are two vacuum chambers, which give a conduction and convection thermal shielding from RT thermal load. In between two vacuum chambers, there is a LN_2 chamber to reduce the thermal gradient from room temperature to 77 K and 77 K to liquid helium chamber. The maximum vacuum with LN_2 and LHe has been achieved up to chambers has 6×10^{-7} mbar. This arrangement provides flexibility for pre-cooling of liquid nitrogen and liquid helium chambers by liquid nitrogen, which reduces consumption of liquid helium during GMS operation for long duration.

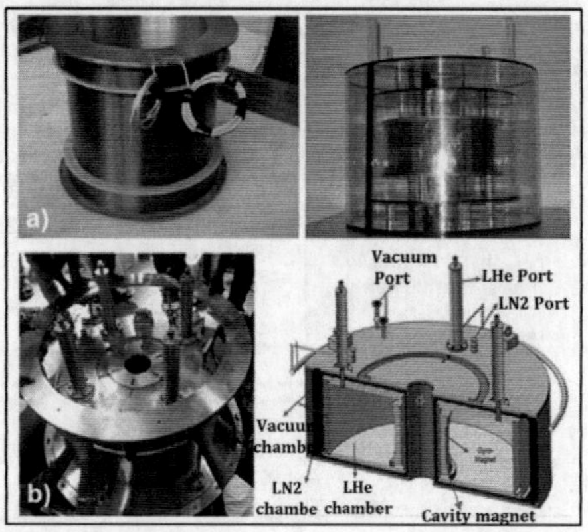

Fig.3.a) Superconducting cavity magnet and its prototype b) Cryostat for cavity magnet

4 Assemblies and Alignment of GMS

The assembly and integration of gyrotron magnet system has been carried out by adopting bottom to top approach [11]-[12]. Gun magnet is assembled on bottom side of top flange of oil tank with suitable supports. The cryostat housing Cryo-magnet is assembled on the top side of oil tank flange maintaining center of the bore hole. The supporting flange for collector magnets sub-assembly is installed at top of ring support flange. The collector magnet sub assembly is then assembled on the supporting flange. The general methodology for assembly built up for GMS with gyrotron is shown in Fig. 3. During assembly, the alignment of magnets has been carried out and measurement for the geometrical axis has been done. The assembly and alignment of GMS is validated with the help of geometrical measurements as well as magnetic field mapping in radial as well as axial direction. The fine tuning for position of various magnets in GMS is done based on experimental measurement of fields. It is ensured that magnet axis and geometric axis has been aligned within acceptable limit ± 2 mm.

Cryostat is provided with number of ports for current lead connection, transfer lines for coolants and thus it is critical to perform leak testing. The cryostat vacuum, LN2 and LHe chamber have been leak tested at room temperature in vacuum mode with gaseous helium with helium leak detector. All the joints and critical areas are monitored for any leaks. Leak rate is observed to be 1.5×10^{-9} mbar l/s which is acceptable [4]. Liquid helium level inside the LHe chamber is continuously monitored by dedicated level sensor (AMI-make). The calibration and validation of liquid level sensor for cavity magnet is completed and reported [13].

DOI: 10.1051/epjconf/201714704005

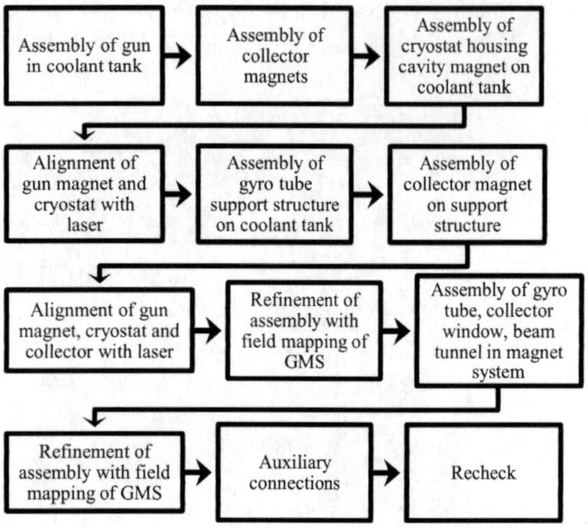

Fig. 4. Typical assembly sequence for GMS

5 Performance Validations of GMS

The alignment of GMS with required accuracy has been done. The validation of its performance has to be establishing for long duration operation to generate the high power electron beam from 42 GHz gyrotron system. It is required to assure that cavity magnet has been superconducting and the voltage across cavity magnet should not exceed quench voltage (250 mV). The cavity magnet was instrumented with different type of sensors for precise monitoring, control and protection during operation of gyrotron. These sensors are LHe level sensor, hall probes, current monitoring system, quench detection and interlocks with power supply. Fig.10 shows functional block diagram of quench detection and inter-lock system. In order to optimize quench detection and protection, level of LHe and magnet quench voltages has been taken as an inter-lock signal to power supply. The superconducting cavity magnet is charged only when two necessary conditions are met.

1) LHe level should be > 10.0 inch during charging.
2) Voltage across magnet including joints and current leads should be < 250 mV.

Voltage and level signals are monitored and extended up-to quench detection system through twisted pair of copper cables over a distance of 5 m- 10 m. The level and current signal is first converted to voltage signal in quench detection system and monitored at DL750 oscilloscope. The detected magnet voltages as well as LHe level are compared to the respective threshold voltage of 250 mV and threshold time of 100 ms to activate interlock signal. All acquired signals are then digitized, monitored and stored with a DL750 scope coder based data acquisition system. The power supply for cavity magnet has been configured to an external quench protection instead of a self-quench protection to avoid false quench trigger to a power supply system. The external quench detection system filters all the spurious high-voltage noise spikes coupled to a magnet system during operation of gyrotron system. The parameters for the quench detection and inter lock are mentioned in Table 5.

Table 5. Parameters for quench detection and interlock system

QD Parameters	Values
Magnet voltage detection	
Voltage threshold	250 mV
Threshold time	100 ms
Measurement accuracy	1 mV, 1 ms
LHe level detection	
Level threshold	10.0 inch
Threshold time	100 ms
Measurement accuracy	0.1 inch, 1 ms

The standalone testing of individual magnets as well as testing of assembled GMS have been done at their respective current values. The inductance and resistance of individual magnet at room temperature has been measured with LCR meter (make-HIOKI 3522-50). It is shown in Table 6.

Table 6. Measured inductance and resistance value of magnets at Room Temperature (RT)

Type of magnet	Inductance (H)	Resistance (Ω)
Cavity	11.76	346
Collector Magnet-1	1.33	12.3
Collector Magnet-2	0.3	5.3
Collector Magnet-3	0.3	5.46
Gun Magnet	0.4	6.2

Dedicated individual power supplies are used for magnets with built in controller to obtain required ramp rates and current profiles. The superconducting cavity magnet is bath cooled in LHe housed inside the low loss cryostat. The performance of superconducting cavity magnet is observed experimentally at various current from 10 - 50 A at r=0 mm along z-direction. In Fig.5, peak field value near interaction region in cavity magnet is uniform from 331 mm to 371 mm along z-direction. The entire length of cryostat bore is 400 mm. With respect to assembly measurements, it starts from z=173 mm and ends at z=573 mm. From this measurement, the center position of cavity magnet has been observed at z= 353 mm, which is under acceptable deviation of ±2 mm in axial direction with respect to desired value z= 351 mm. The magnetic field profile for cavity magnet has been evaluated from bottom of gun magnet to top of the cryostat. The homogeneous and uniform magnetic field has been observed for ± 25 mm in radial (r) direction and ± 20 mm in axial (z) direction at electron beam interaction region. The helium liquid level sensor with its monitor is also connected for monitoring of liquid level in LHe chamber. After cool down of cavity magnet, it has been energized up to 50 A with ramp rate of 0.02 A/s. During charging of cavity magnet, liquid level of helium has been maintained at 10.9 inch - 11.0 inch.

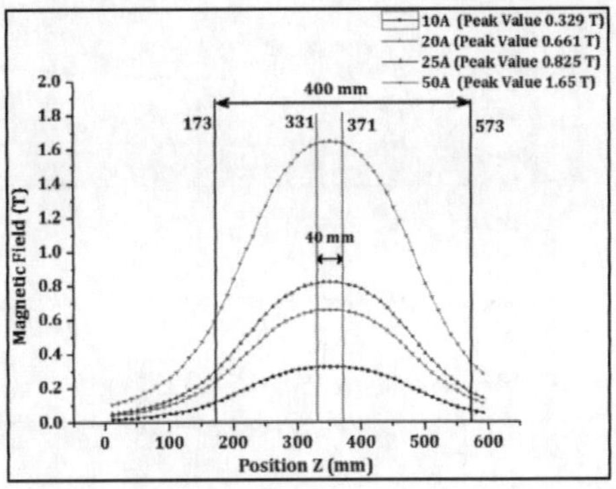

Fig. 5. Field profile of cavity magnet at various current

Transfer of LHe, filling of LN2 filling in 77 K shield, charging of cavity magnet, charging of resistive magnets, maintaining the liquid level and vacuum level within the cryostat during test of GMS have been shown in Fig.6. The vacuum level has been shown in Fig. 6 (b), (c) and (d), before LHe transfer, after LHe filled and vacuum valve closed respectively. The vacuum level has got changed to 5.7 × 10^{-7} mbar. Magnetic field mapping has been done by energizing all magnets of GMS at their respective nominal currents. The variation of magnetic field profile in axial and radial directions has been shown in Fig.7. The field is observed to be homogeneous for ± 20 mm in axial direction at beam interaction region and ± 25 mm in radial direction throughout the length of gyro-tube. It is observed to be as per the designed requirements for optimum performance of gyrotron. The variation in peak field values near center region of cavity magnet has been measured below ±1%. This result shows that all individual magnets in GMS are aligned as per technical requirements for optimal gyrotron operation. The custom made setup having gauss meter (make- FW Bell) has been fabricated to measure magnetic field along the z-axis at r =0 mm and r =±24 mm within bore of GMS. An experimental measurement of magnetic field profile of GMS has shown in inset of Fig.8. Small deviation (within ±1%) has been observed in peak field value near center of cavity or at interaction of electron beam. This deviation is acceptable with respect to designed tolerance. All magnets have been positioned and aligned successfully as per the assembly drawings of 42 ± 0.05 GHz gyrotron system.

Fig. 6.a) LHe transfer for cavity magnet b) Vacuum at RT c) Vacuum during LHe transfer d) Vacuum when Vacuum Valve closed

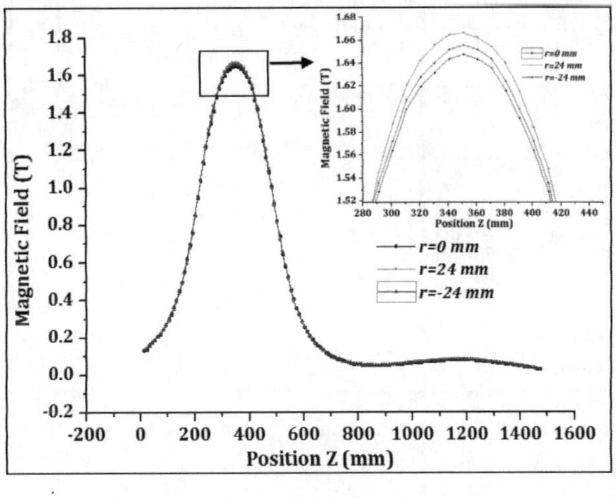

Fig.7. Magnetic field profile along the z- direction

DOI: 10.1051/epjconf/201714704005

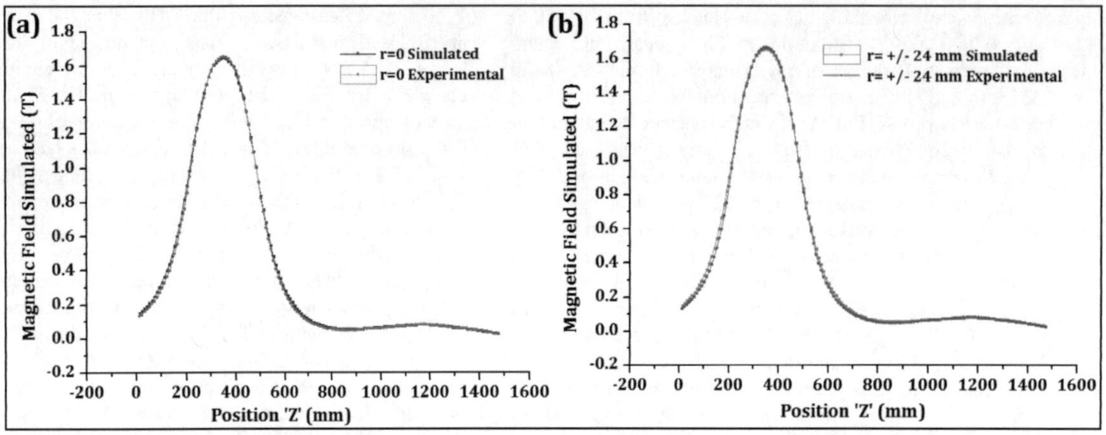

Fig.8. Simulated and experimental field profile at r=0 mm and r = ± 24 mm

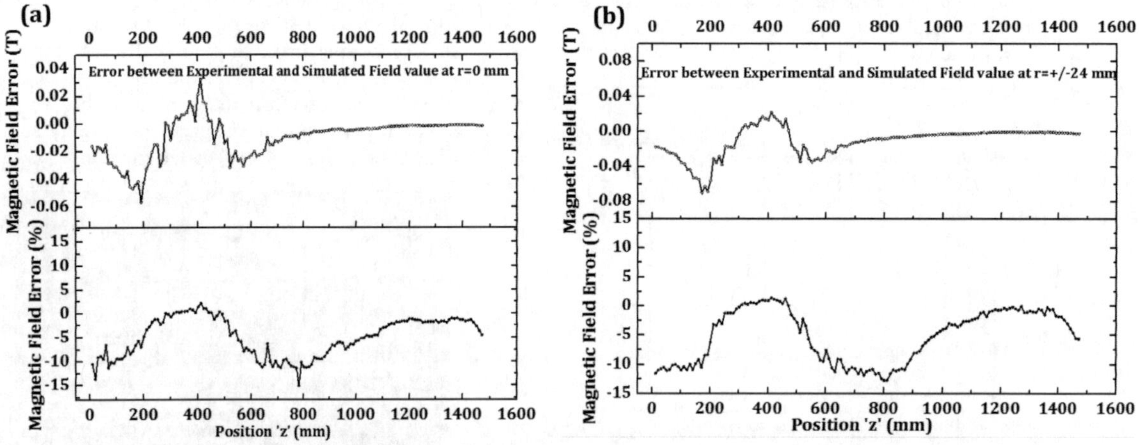

Fig.9. Error between experimental and simulated values at r=0 mm and r = ± 24 mm

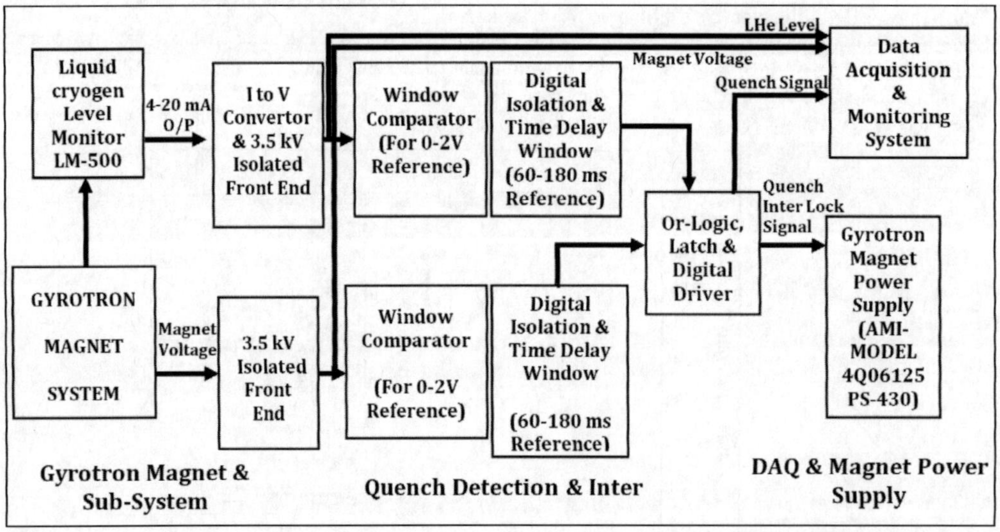

Fig.10. Functional block diagram of quench detection and interlock system

The experimental values also have been compared with simulated values of assembled GMS with its FEM (Fig. 8 (a), (b)). Fig. 8 shows comparison of field profile along z-direction at r=0 and r = ±24 mm. Fig. 9 shows error and error percentage at each individual axial position at r=0 and r=±24 mm. The error in magnetic field throughout the z-locations in bore of GMS has been shown in Fig, 9. In Fig. 9 (a) and (b), it is observed that the magnetic field error near beam interaction region is well below ± 1% and within ± 5% at other locations of z. The measured error field in the interaction region is ~0.3-0.7 %. As per design it should be below 1 %. The magnetic field was measured with Hall probe located at Z= ~554 mm and R = ~80 mm. At this location, the field will be around 0.3 kG at 48 A, which corresponds to 1.5 T at the cavity coil center, as shown in figure 13. Similar measurement at lower current up to 38 A is also shown in figure 11. The variation in peak magnetic field has been observed near beam interaction region ± 0.01 T, which is under the acceptable limit.

6 Operation of GMS during test

This configuration has been successfully tested and verified with multiple charging of GMS during gyrotron tube operation. The magnetic field and transport current of cavity magnet were measured with hall probe, voltage across 100 mΩ shunt resistor and respective precision electronics. The hall probe is an active sensor which operates with control current of 100 mA. Fig.11 shows the performances of electronics on measurement of transport current, magnet voltage and magnetic field at different charging conditions of the cavity magnet.

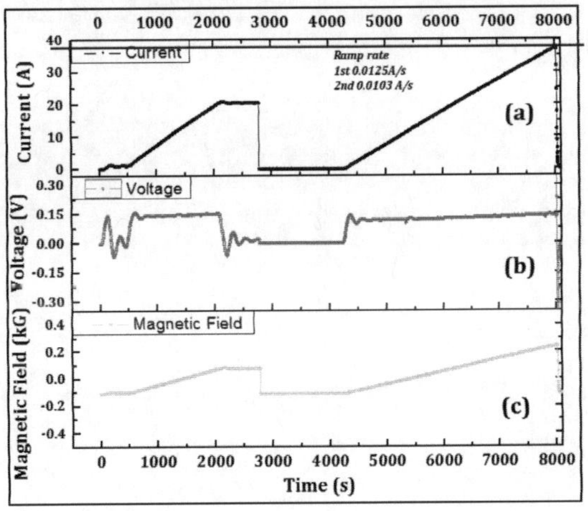

Fig.11. (a) Transport current, (b) Magnet voltage and (c) Magnetic field at different charging conditions of the cavity magnet.

Fig.12 shows the voltage rise with respect to current confirming superconducting behavior of cavity magnet. Slow ramp rate of 0.012 A/s has been chosen to confirm the superconductivity of cavity magnet. The maximum voltage drop during charging with ramp rate of 0.012 A/s was found to be 141 mV. In Fig.13 negligible voltage fluctuation has been observed. The cavity magnet has also been charged with 0.0209 A/s ramp rate. The quench circuit functions as designed with the cavity magnet for long duration charging of magnets. Fig. 13 shows successful long pulse operation (~ 4.5 Hrs.) with charged cavity magnet without any interruption during one of the several experimental operations of gyrotron system.

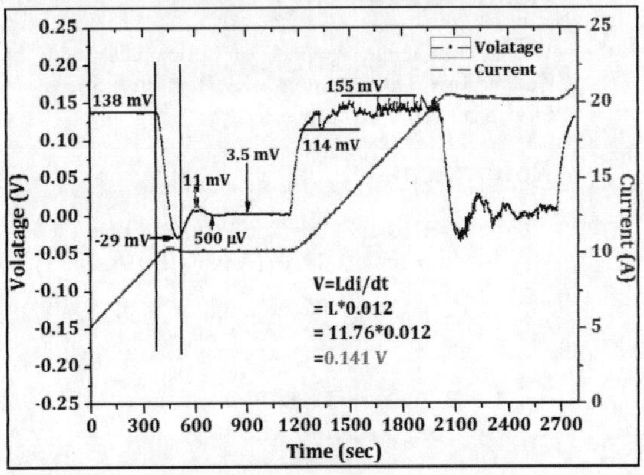

Fig.12. Current ramping in cavity magnet

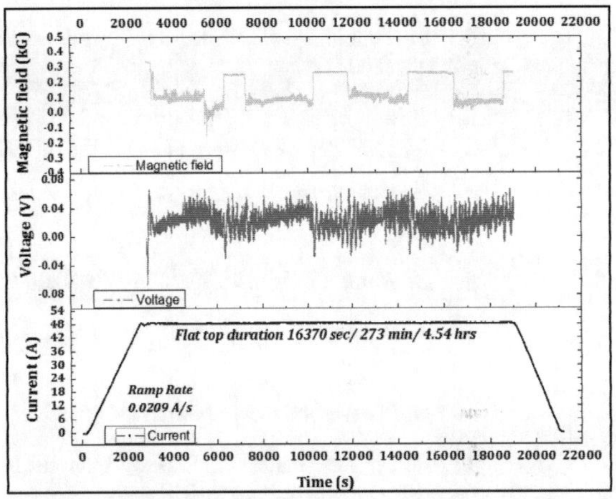

Fig.13. long duration operation of GMS

7 Conclusion

Design of appropriate magnet system comprising of resistive and superconducting cavity magnet considering physics requirements of beam have been experimentally validated. The manufacturing of collector magnet and gun magnet using enameled copper conductor has been successfully achieved. Superconducting cavity magnet in Helmholtz configuration has been manufactured using NbTi superconducting strands and encased in low loss cryostat. The performance validation of individual magnet with respect to design is carried out by experimental measurements at their operating conditions. The assembly of all magnets in GMS is completed using bottom to top approach. The refinement and optimization of assembled magnets is done by field mapping in axial direction. It is observed that field variation is within ±1% tolerance in z-direction and field profile at radial position of r = 0 and r = ± 24 mm. It confirms the uniformity of field along the z-direction. GMS has been found to be suitable as well as reliable for its long duration operation to support high energy electron beam generation in gyrotron. The operation of gyrotron magnet system for long hours has been demonstrated

Acknowledgment

We acknowledge support of colleagues from Magnet Technology Development Division and SST-1 Magnet Division during this project.

References

1. S. Kedia, S. Pradhan, *IEEE* Trans. on App. Supercond., **20**, 4, (August 2010), 2235.
2. S. Kedia, S. Pradhan, Int. Res. Report, IPR (200 8)
3. R. Prakash, DST/IPR/Gyro/12-13/ASM/0, Int. Tech. Report, IPR.
4. J. P. Anderson, M. A. Shapiro, R. J. Temkin, I. Mastovsky, and S. Cauffman, (June 2004).
5. M. Thumm, *Fusion Engg. & Design,* **66-68**, 69-90, (2003).
6. G.B. Ferguson, N.W. Kerley and G.A. Kirby *Fusion Engg. & Design,* **20**, 311-316, (1993).
7. S. Sanz, *IEEE* Trans. on App. Supercond., (July 2007).
8. Q. Wang, *Cryogenics,* **47**, 364–379, (2007).
9. X. Donghui, Plasma Sci. and Tech., **16** (4), (April 2014).
10. D. Bhavsar, Internal Report, IPR.
11. P.Raj, M. Ghate, U. Prasad, Internal Report, IPR.
12. P.Raj, U. Prasad, M. Ghate, *IEEE* Trans. on App. Supercond., **26,** 8, (December 2016), 4902607.
13. M.Banaudha, Internal Technical Report, IPR.

EPJ Web of Conferences **147**, 04006 (2017) DOI: 10.1051/epjconf/201714704006
EC-19

Design and Research of Electron Cyclotron Resonance Heating and Current Dive System on HL-2M Tokamak

Mei Huang[1,*], *Jao* Rao[1], *Shaodong* Song[1], *He* Wang[1], *Gangyu* Chen[1], *Zhihua* Kang[1], *Feng* Zhang[1], *Jieqiong* Wang[1], *Jiruo* Ye[1], *Kun* Feng[1], *Bo* Lu[1], *Bo* Huang[1], *Chao* Wang[1], *Mingwei* Wang[1], *Weimin* Xuan[1], *Lieying* Rao[1], *Qing* Li[1], *Xiaohui* Mao[1], *Xuru* Duan[1], *Yong* Liu[1]

[1]Southwestern Institute of Physics, P.O. Box 432, Chengdu, China

Abstract. A research has been conducted to develop an 8MW electron cyclotron resonance heating and current drive (ECRH/ECCD) system on HL-2M tokamak. The ECRH system compromise eight 1MW gyrotrons, eight evacuated transmission lines and three launchers. The main purpose of the ECRH system was to suppress the neo-classical tearing modes and control the plasma profile. This paper presents an overview of the design and studies performed in this framework. Some primary test results of the critical components have been released in this paper, e.g. polarizers, power monitor and fast steering launchers.

1 Introduction

Electron cyclotron Resonance Heating and Current Drive (ECRH/ECCD) system is employed worldwide for plasma heating and control on fusion research devices, with the merit of remote steering, localized heating and current drive etc.[1-3] ECRH also played a very important role on HL-2A tokamak during the last decade, such as plasma heating, MHD instability control, assisted start-up, real-time TM suppression, transport study, and so on.[4-9] In the next several years, the HL-2A tokamak will be upgraded to HL-2M, which has larger vacuum vessel and higher performance of plasma.[10] The major radius of HL-2M will be 1.78m and minor radius 0.65m. The maximum toroidal magnetic is expected to be 2.2T and plasma current is about 2.5MA. For the HL-2M tokamak, ECRH/ECCD system will also be acted as one of the key plasma heating method for central electron heating, current profile control and NTM suppression.

Since the HL-2M will be operated at 1.8T normally and the biggest toroidal magnet field will up to 2.2T, the frequency of ECRH system is fixed at 105GHz and 140GHz. The ECRH system on HL-2M tokamak was designed to compromise eight 1MW gyrotrons (two of them are dual-frequency, the others are single frequency 105GHz), eight evacuated transmission lines and three steerable launchers. Fig. 1 shows the sketch of the layout of ECRH system. All of the gyrotrons will be settled in the RF heating hall which is in the south of HL-2M tokamak hall. Three ports in the east of HL-2M, one equatorial port and two upper ports situated at the same vacuum segment, were allocated for ECRH launchers. For the launchers, care is principally taken to change the wave injection angle in poloidal direction in real time. A preliminary design of the transmission line routing has been performed, mainly focused on getting the best solutions of attenuation issues. The first plasma of HL-2M was expected to be discharged at the middle of 2018. By then, 5MW EC power could be delivered into plasma according to the process schedule of the ECRH project.

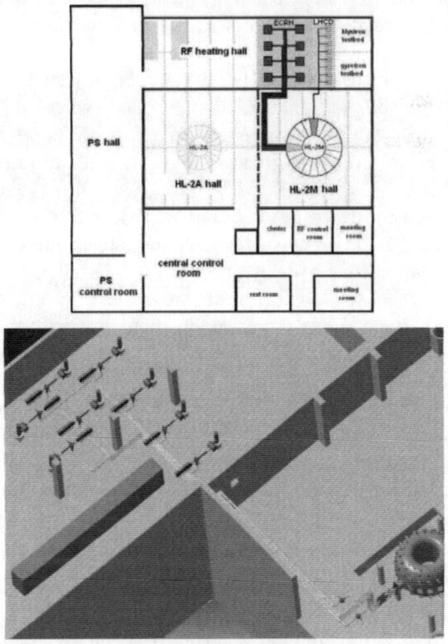

Fig. 1. Sketch of the layout of ECRH system and waveguide routing from the RF heating hall to the HL-2M tokamak hall

* Corresponding author: hm@swip.ac.cn

2 Gyrotron commissioning

All of the 1MW eight gyrotrons for HL-2M ECRH system are GYCOM gyrotrons. In order to commissioning the gyrotron flexibly, an independent test bed has been build last year in the RF heating hall, shown in Fig. 2. This test bed includes two different calorimeter dummy loads and body power supplies, one for short pulse and another for long pulse.

Fig. 2. Test stand for gyrotron commissioning

Based on the operation experience on HL-2A for many years, the basic control system was also constructed around PLC for the test bed and HL-2M ECRH system. The timing of gyrotron pulses as well as the launcher rotation angle could be controlled during the plasma discharge by fast control system. The interlock and interrupt circuits were taken into account to operate the ECRH system in safety for man and device. Same as the HL-2A ECRH system, Pulse Step Modulation (PSM) technology was adopted for the body HVPS of gyrotron which was fed by freewheel motor-generator and four gyrotrons shared one HVPS.[11] The output voltage of body HVPS could be adjusted from -20kV to -80kV, ripple less than 1% and switch-off time less than 10μs. For convenience, a new body PSM HVPS only for one gyrotron and powered by intercity network was developed to fit for the operation requirement of the test bed. Further research of the body PSM HVPS are being carried out to increase the voltage modulation frequency up to 5kHz for NTM suppression. Simultaneous modulation of the collector voltage is necessary for deep power modulation. A new collector HVPS based on PSM technology was developed to achieve this aim. Tested with a resistor load, the maximum modulation frequency of it could reach 5kHz. At the beginning of this year, the commissioning of two 105GHz gyrotrons has been fulfilled on this test bed. The hot test of transmission components will be put into effect on the test bed this year.

3 Transmission line

For the consideration of space limit and safety, a kind of evacuated transmission system carrying HE_{11} mode with low loss, high power capability, high transmission efficiency and broad bandwidth is designed and manufactured for HL-2M ECRH system based on the over-moded circular corrugated waveguide theory.[12-14] The eight 1MW wave beams will be transported by eight evacuated transmission lines and be injected into the plasma through three launchers. The scheme of the eight transmission lines is shown in Fig. 3. Six of them are connected with the equatorial launcher, another two jointed with upper launchers. The lines will be about on an average 40 meters in length, including 63.5mm waveguides, six or seven miter bends, bellow, DC break, pumping tee, switch and calorimeter dummy load. For the transmission system, some key issues have been considered, e.g. power attenuation, wave polarization and power measurement. At present, design and fabrication of all these transmission components have been finished.

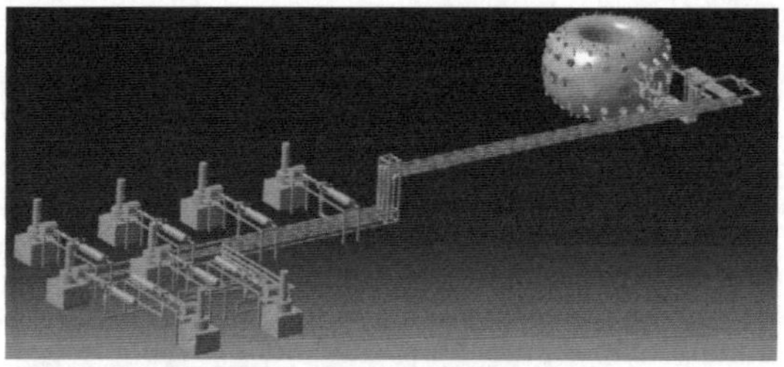

Fig. 3. Layout of transmission line for HL-2M ECRH system

3.1 Attenuation analysis of TL

Table 1. Loss of transmission line at different frequency

Loss		140 GHz	105GHz
Ohm (dB)	~ 30m corrugated WG	$2.58e^{-3}$	$3.99e^{-3}$
	4 flat mirrors	$2.56e^{-2}$ ~ $5.08e^{-2}$	$2.20e^{-2}$ ~ $4.40e^{-2}$
	2 polarizers	$2.56e^{-2}$ ~ $5.12e^{-2}$	2.20e-2 ~ $4.40e^{-2}$
Coupling (dB)	TEM_{00} to HE_{11}	$8.8e^{-2}$	$8.8e^{-2}$
	0.2^0 tilt	$1.36e^{-2}$	$0.77e^{-2}$
	2 mm axial offset	$2.08e^{-2}$	$2.08e^{-2}$
Misalignment (dB)	0.2^0 tilt	$1.43e^{-2}$	$0.77e^{-2}$
	2 mm axial offset	$2.48e^{-2}$	$2.48e^{-2}$
Gap between WG(dB)	6 miter bends	0.166	0.18
	~ 10mm other gaps	$4.7e^{-3}$	$7.2e^{-3}$
Total (dB)		0.386 ~ 0.44 (6.9% ~ 7.8%)	0.384 ~ 0.428 (7% ~ 7.7%)

The losses at different frequency, 105GHz and 140GHz, in the transmission line have been estimated theoretically. The losses are mainly caused by ohmic attenuation and mode conversion and the mode conversion losses are mainly caused by coupling, misalignment and gaps.[15-16] Normally, ohmic attenuation is inherent and very low. And the attenuation of mode conversion are the key parts. The analysis results are presented in Table 2. It can be seen that the transmission efficiency is about 92%.

3.2 Wave polarization control

Two polarizers will be integrated on the miter bends to control the wave polarization. The polarizers could be rotated fast.

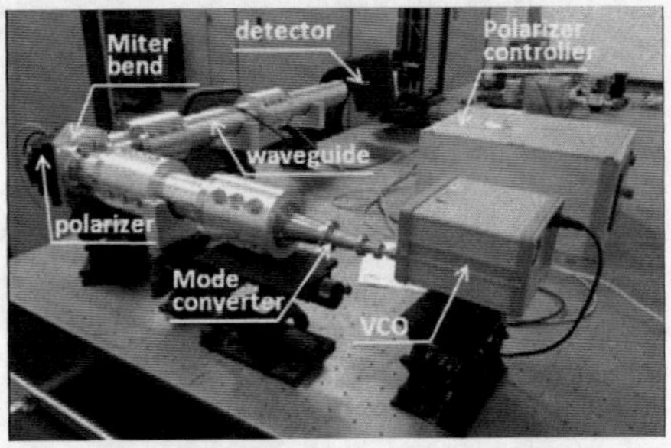

Fig. 4 Fast rotatable polarizer and laboratory test bed for polarization measurement of polarizers

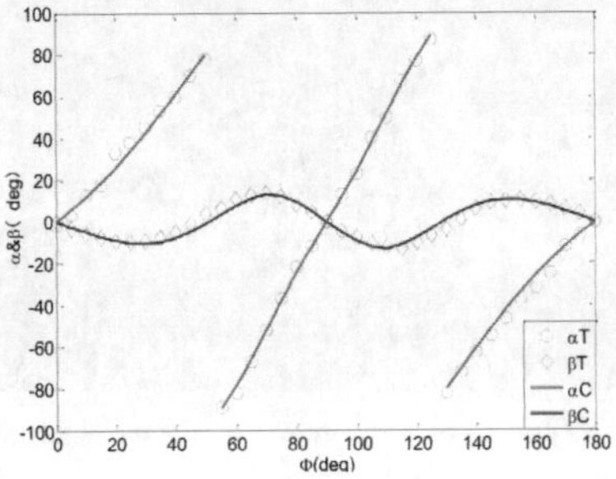

(a)

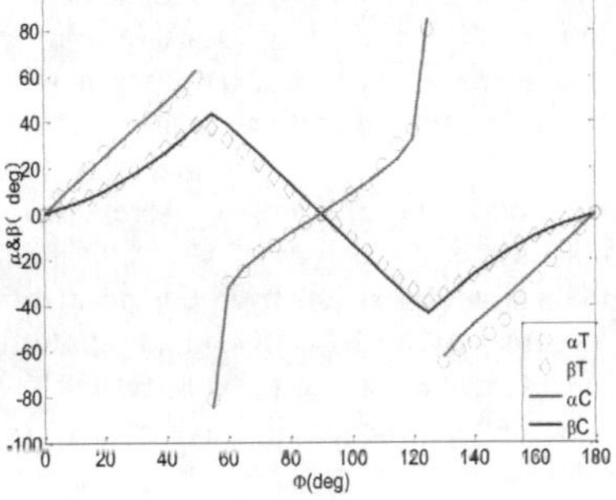

(b)

Fig. 5 Polarization parameters of the theoretical simulation results and cold test results for (a) the linear polarizer and (b) the elliptical polarizer, αC and βC represent the calculation results, αT and βT are for the low power test results, Φ for polarizer rotation angle

The two optimized polarizers are designed for single frequency by a simple and rigorous way.[17] Arbitrary polarization of wave could be obtained by rotating these two polarizers. So the desirable mode could be excited for high coupling efficiency between EC waves and plasma when the millimeter waves injected into plasma obliquely. The cold test of these two polarizers have been explored and the experimental results are consistent with the theoretical simulation results nicely which illustrate that both the design and test method are reasonable.[18] The laboratory test bed for measurement the ellipticity β and axis rotation angle α is shown in Fig.4 and test results are shown in Fig.5. A study of polarization control strategy which are suitable for double frequency, 105GHz and 140GHz, are on-going, makes it unnecessary to replace the polarizers in the transmission lines when the gyrotrons operated at different frequency.

3.3 Power measurement

Two power monitors are planned to be settled on the miter bends to measure the wave power. One is located near the gyrotron and another near the launcher, not only to estimate the transmission efficiency but also to follow the operation status of the gyrotrons. In order to validate the feasibility of the design and machining process, a 68GHz power monitor for HL-2A ECRH system was manufactured and checked during the plasma heating experiments at first, shown in Fig. 6.

Fig. 6 Directional coupler settle on miter bend and calibrated by calorimeter dummy load

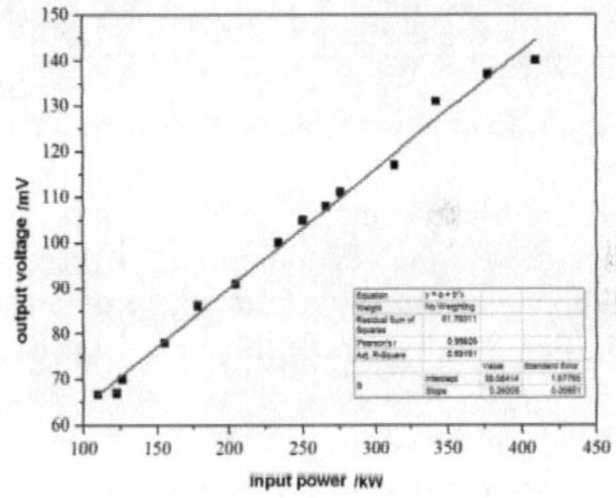

(a)

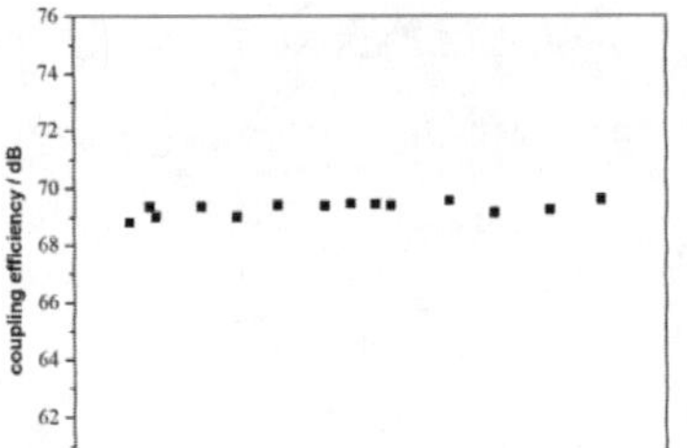

(b)

Fig. 7 (a) Dependence of output voltage of detector on input power of directional coupler and (b) the coupling coefficient of directional coupler: -69.3dB

The power monitor has been calibrated by the calorimeter dummy load. It can been seen from the Fig. 7(b) that coupling coefficient was about -69.3dB at different wave power. Fig. 7(a) shows the dependence of the output voltage of detector on the power measured in the dummy load. wave power and the linearity between input power and coupling output power is perfect. An 105GHz monitor has been produced in the same way as 68GHz monitor and the high power test results will be released in this year. A research for directional coupler with two-row-hole which could measure the wave power and wave polarization at the same time is proceeding.

4 Launcher

The three launchers of HL-2M ECRH system are one Equatorial launcher(EL) for 6MW, one upper launcher(1#UL) for 2MW and one upper launcher(2#UL) for 1MW. They have different functions, shown in Table 2.

Table 2. Performance of launchers for HL-2M ECRH system

	EL	1# UL	2# UL
Port Size	450x500	Ø300	Ø200
Scheme	3x3	1x2	1x1
Frequency	105GHz	105/140GHz	105GHz
Power	6MW	2MW	1MW
Pulse	3s	3s	3s
Scan range	Toroidal: -20~20 Poloidal: -20~14	Poloidal: -80~-15	Poloidal:-90~-30
Function	Plasma heating, current drive, assisted startup	NTM suppression	NTM suppression

The main functions of EL is plasma heating, current drive and assisted start-up. The purposes of two upper launchers are MHD control and NTM suppression. The EC power deposition location almost covers the whole cross section of plasma by rotating these three launchers, shown in Fig. 8. All of these three EC launchers enable millimeter-wave beam continuous scanning in poloidal directions in real time.

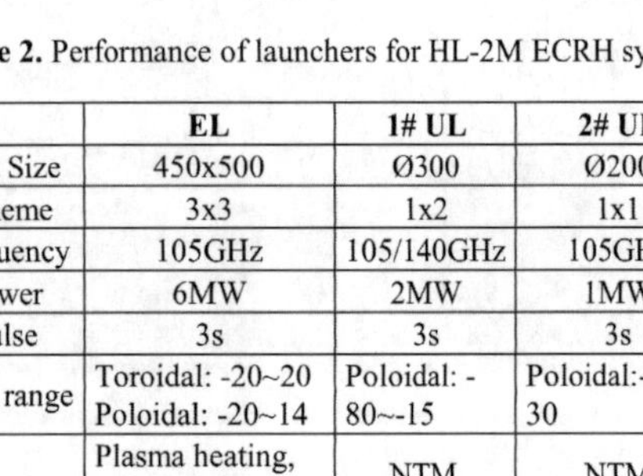

Fig. 8. Scan range of equatorial launcher and 1# upper launcher on the cross section of plasma

4.1 Equatorial launcher

The 6MW EL has been fabricated and tested at the end of 2015, which could change the ECW injection angle at two-dimension. The outer structure of the plug-in unit and equatorial launcher is shown in Fig. 6.

Fig. 9. Outer structure of the plug-in unit and equatorial launcher

Available space, port shape and port size of HL-2M tokamak suggest using two different kinds of steering mechanism for the poloidal and toroidal movement. In order to save space, wire rope and servo motor were applied in the toroidal direction in that the structure is simple and tight. In the poloidal direction, more complicated push rod framework were employed for NTM suppression in real time. A faster Kollmorgen linear motor was chosen to drive the push rods which can be controlled by NI-cRIO main controller. The performance of the EL has been tested. The total response time of control activities is less than 100μs and the dynamic response time of mechanism for the full scan range is 45ms in the poloidal direction, which can fully satisfy the requirement of changing the ECW power deposition during one plasma shot.

4.2 Upper launcher

The preliminary design of 1# upper launcher has been performed and the layout is shown in Fig. 10. Compared with the EL, the space for housing the reflective mirrors and steering mechanism is not enough because of the port size and shape. An extra extended segment was designed to solve this problem which caused the push rod becoming longer.

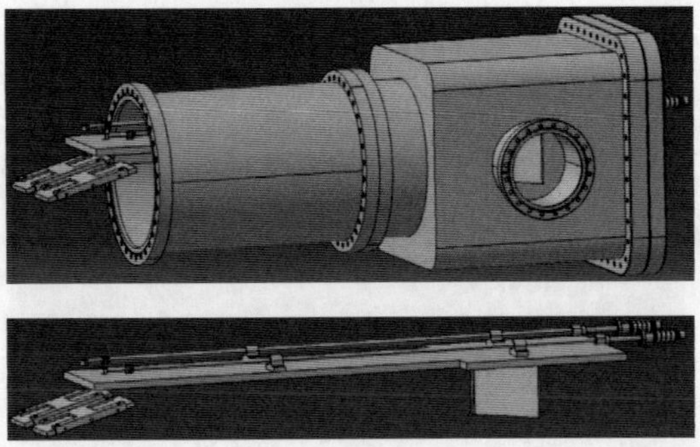

Fig. 10. Preliminary design of 1# upper launcher

A table test has been done to testify the performance of the steering mechanism. A prototype of the real steering mechanism and flat mirror were manufactured and assembled on the optical platform. Motor with higher qualities and the same main controller as EL were adopted in this case. The steering mechanism moved once every five minutes, up to one hundred times tests have been finished in a day. The maximum speed of motor reached 6000rpm and the mechanical structure maintained good condition. When the acceleration speed was set as 290000rpm/s, the response time of the movement for full range was 45ms. The test results show that the performance of 1# launcher could totally meet the demand of real time NTM suppression on HL-2M tokamak. The fabrication of 1# UL will be completed at the end of this year.

5 Conclusion

An 8MW ECRH system is expected to be constructed step by step on HL-2M. The conceptual design of this system has been undertaken. Most critical issues concerning transmission system and launchers have been studied and solved. The key components, such as PSM PS, fast rotable polarizer, power monitor and mirror steering mechanism, have been developed and tested. From the technical point of view, the behavior of these components could meet the physical objectives of HL-2M tokamak.

ACKNOWLEDGMENTS

Project is partly supported by the International Thermonuclear Experimental Reactor Special Fund of China (Grant No. 2013GB103000) and the National Natural Science Foundation of China (Grant No. 11175059). The authors appreciate all the staff and researchers contributing to the ECRH system on HL-2M and are grateful to the specialists of GYCOM, S. Kubo (NIFS, Japan) for fruitful discussion and suggestion.

References

1. V. Erckmann, U. Gasparino, et al., Plasma Phys. Control. Fusion **36**, 1869 (1994)
2. B. Llyod, et al., Plasma Phys. Control. Fusion **40**, A119 (1998)
3. M. Thumm. et al., 2003, Plasma Physics and Controlled Fusion, **45**:A143
4. X.R. Duan, et al., Nucl. Fusion **50**, 095011 (2010)
5. X.T. Ding, et al., Nucl. Fusion, **53**, 043015 (2013)
6. W.L. Zhong, et al., Nucl. Fusion **53**, 083030 (2013)
7. X.Q. Ji, et al., Chinese Physics Letters, 27, 6(2010)
8. X. Song, et al., IEEE Transactions on Plasma Science, 42, 3(2014)
9. M. Huang, et al., 3MW ECRH system and experiments on HL-2A, 37th EPS conference, Dublin, Ireland (2010)
10. Dequan Liu, et al., Fusion engineering and design, **96-97**, 298-302(2015)
11. Xiaohui Mao, et al., IEEE Transactions on Plasma Science, **42**, 1425-1429(2014)
12. M. Thumm, et al., Fusion Eng. Des., **26**, 291 (1995)
13. Doane John L., et al., Propagation and mode coupling in corrugated and smooth-wall circular waveguides. Presented in Infrared and Millimeter Waves, Button K J, ED. New York: Academic, **13**, 123 (1985)
14. Dragone C. Bell Syst. Tech. J., 56, 835(1977)
15. K. Ohkubo, et al., Int. J. Infrared Millim. Waves, **18**, 23(1997)
16. Doane John L., et al., Int. J. Electron, 77, 489(1994)
17. D.H.Xia, et al., Review of scientific instruments, **84**, 103504-103504(2014)
18. F. Zhang, et al., Journal of Infrared, Millimeter, and Terahertz Waves, **37**, 572-581 (2016)

DOI: 10.1051/epjconf/201714704007

Design and Analysis of Steerable ECRH Launcher for SST-1 Tokamak

Hardik Mistry*, and *B K* Shukla

Institute for Plasma Research, Bhat, Gandhinagar, Gujarat, India -382428

Abstract. In the tokamaks ECRH system is used for pre-ionization, start up, heating, current drive and suppression of NTMs (Neo Classical Tearing Modes). A Standard ECRH system consists of high power microwave source Gyrotron, circular corrugated waveguide based transmission line and launcher. The Focused ECH power is launched into plasma through launcher. The microwave beam emerges out from circular corrugated waveguide and propagates freely in air with finite divergence. So focusing and plane mirror combination is used to launch focused beam in plasma. Thus an ECRH launcher consists of metallic profiled and plane mirror, UHV compatible vacuum barrier window and a UHV gate valve. One 42 GHz gyrotron capable of delivering 500 kW of power for 500 ms and other 82 GHz gyrotron capable of delivering 200 kW of power for 1000s are used for SST-1 ECRH system. The launcher design consists of mirror design, design of supports and design of steering mechanism to provide suitable movements with minimum backless error. The whole assembly is UHV compatible. The launcher is capable of steering the beams by ±20° in both toroidal and poloidal directions. Mirrors are given motion by means of one rotary and one linear feedthrough. For 82 GHz launcher active cooling is provided, whereas for 42 GHz launcher no active cooling is provided.

A detailed analysis is carried out for the mirrors of the high power launcher. The heat load for the 82 GHz launcher is 2 kW (~1% absorption) and for 42 GHz launcher it is 5 kW. For 82 GHz launcher, the maximum steady state surface temperatures of focusing and reflecting mirrors are 315K and 323K and von-mises stresses are within 10 MPa. Similarly for 42 GHz launcher maximum temperatures observed during 500 ms pulse are 301K and 303K for focusing and reflecting mirrors respectively. This paper explains the mechanical and thermal design and analysis of the launcher for the ECRH system.

1 Introduction

The ECRH offers the advantages of high coupling efficiency and fine control over power deposition. So, it is one of the attractive options for heating of tokamak plasma and it has been employed for many tokamaks. In SST-1, along with the ICRH (Ion Cyclotron Resonance Heating) and LHCD (Lower Hybrid Current Drive), the ECRH is used for pre-ionization, start up, heating and partially driving the current [1]. This system has two microwave sources, one 42 GHz gyrotron capable of providing 500 kW power for 500 ms and one 82 GHz gyrotrons capable of delivering 200 kW power for 1000s. This power is transmitted to tokamak through corrugated waveguide based transmission system which consists of matching optical unit (MOU), DC breaks, meter bends with bidirectional coupler, polarizer and vacuum window [2-3].

The microwave beam emerges out from the corrugated transmission line in TEM_{00} mode which has Gaussian profile and propagates in the vacuum with the finite divergence. In order to achieve desired beam radius at the plasma center, optimization is required. During the pulse, small part of the power (~ 1%) is absorbed by the mirror material and hence its temperature increases. In order to avoid the excessive heating of the mirror, the maximum temperature during the ECRH pulse has to be estimated and active cooling has to be provided as per the requirement.

2 Structure of the launcher

The launcher consists of two metallic profiled mirrors along with its support structure, steering mechanisms, UHV gate valves, vacuum barrier window and cooling arrangement. The mirror assembly of the launcher is shown in the figure 1.Two sets of focusing and reflecting (steering) mirrors are used for 42 and 82 GHz beams. The mirror assembly along with the cover plate is mounted on one equatorial port of the vacuum vessel. The focused beam then incidents on

* Corresponding author: hardik.mistry@ipr.res.in

the reflecting mirror (RM) which is parallel to the focusing mirror (FM). The RM steers the beam in two directions – poloidal and toroidal. The optimized dimensions of the focusing mirror for 42 GHz beam is 215 × 305 mm and for 82 GHz it is 125 × 175 mm.

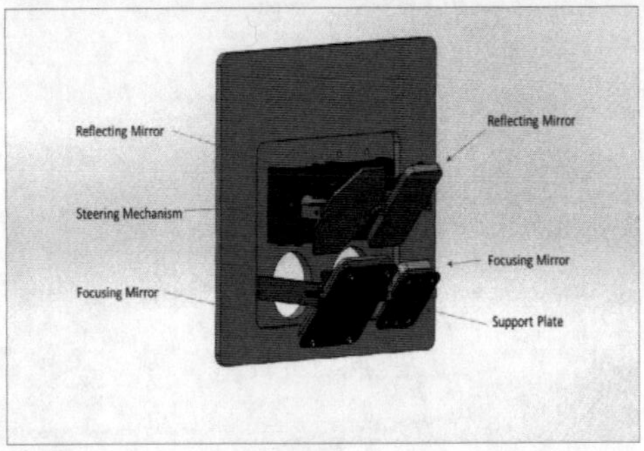

Fig 1. Model of the Launcher

The dimensions of the mirrors are optimized in such a way that more than 99% of the Gaussian beam power incidents on the mirror. Both the reflecting mirrors are connected to UHV compatible motion feedthroughs which facilitate the mirror to move along the poloidal and toroidal directions.

3 Beam optics and RF considerations

The HE_{11} wave when emerges out of the corrugated waveguide it travels in the free space with finite divergence and hence needs to be focused [4]. The beam emerging out of the waveguide has a Gaussian profile i.e. the intensity is maximum at the center and it falls rapidly as the distance from center increases. The intensity (I) and divergence (θ) of the beam can be written as

$$I = I_0 e^{-\left(\frac{x^2}{a_0^2}\right)} \quad (1)$$

and

$$\theta = \frac{1}{k a_0} \quad (2)$$

where,

I_o = peak intensity
x = radial distance
a_0 = beam radius at which intensity falls to its 1/e value
k = wave propagation vector.

The angle θ is the characteristic divergence of a Gaussian beam for the beam waist a_0. The divergence of a Gaussian beam transmitted in space is shown in the figure 3. The general expressions for the beam radius and the radius of the wave front are given by

$$a(z) = a_0 \sqrt{\left[\left(1+\frac{z}{R_0}\right)^2 + \left(\frac{z}{k a_0^2}\right)^2\right]} \quad (3)$$

and

$$R(z) = z \frac{1+\left(\frac{k a_0^2}{z}\right)^2 \left(1+\frac{z}{R_0}\right)}{1+\left(\frac{k a_0^2}{z}\right)^2 \left(\frac{z}{R_0}\right)\left(1+\frac{z}{R_0}\right)} \quad (4)$$

Where a_0 and R_0 are the beam radius and the radius of the incident wave front at z = 0.

The microwave beam emerging from the circular corrugated waveguide in the HE_{11} mode is Gaussian in nature. At the exit of the waveguide, the radius of the wave front $R_0=\infty$.

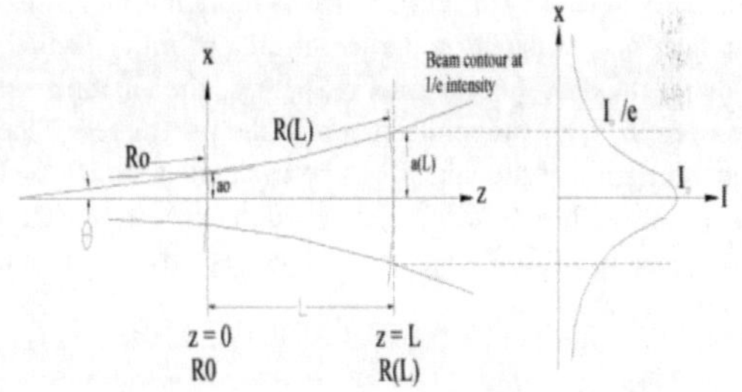

Fig 2. Beam divergence and intensity profile

The schematic diagram of the launcher system is shown in the Fig 3. Here, D_0 is the distance between waveguide end and focusing mirror, D_v is the vertical distance between two mirrors and D_p is the distance between reflecting mirror and plasma center. Here, axial distance between waveguide end and plasma center ($D_0 + D_p$) is fixed for the given layout. Positions of the mirrors can be varied within certain range. However, as the beam comes out of the waveguide it starts diverging.

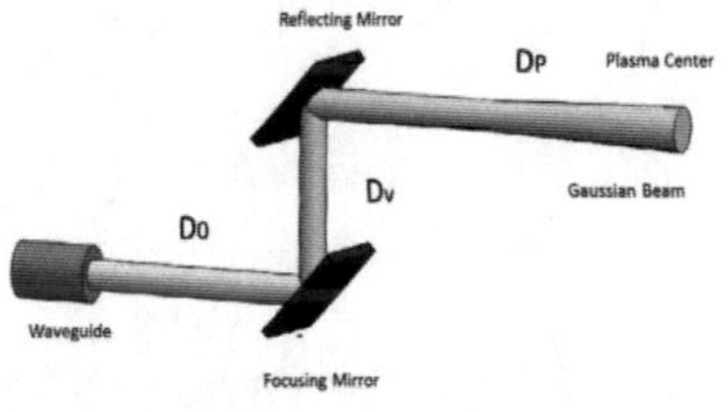

Fig 3. Schematic of Launcher system

DOI: 10.1051/epjconf/201714704007

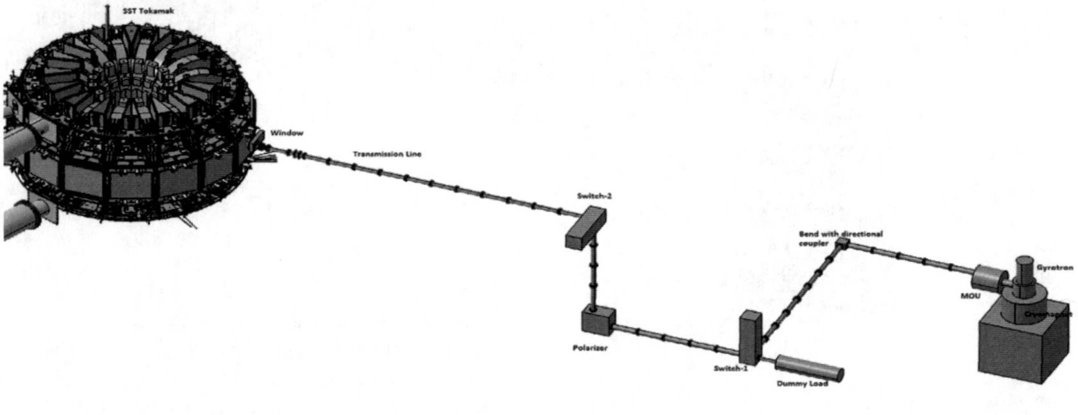

Fig 4. Present layout of ECRH system

So, if distance D_0 is increased the required mirror size will also increase. However, as the beam radius increases, the heat is dissipated on the larger area of mirror which reduces the heat flux for mirror. So, these dimensions can be optimized by proper selection of the parameters. The focal length f of the mirror can be calculated in terms of the radius of the curvature of the incident wavefront R and the reflected wavefront R', which is given in the following expression:

$$\frac{1}{f} = \frac{1}{R} - \frac{1}{R'} \tag{5}$$

In the launcher, the optimized beam radius at the plasma center is 25 mm for 42 GHz system and 20 mm for 82 GHz system. The focal length of the focusing mirrors are 0.41m and 0.54 m respectively.

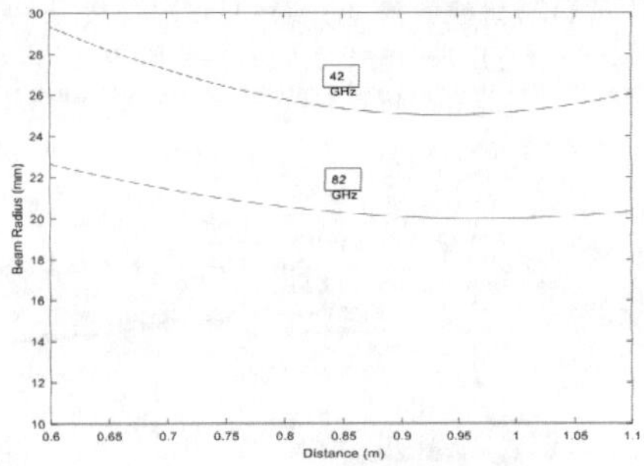

Fig 5. Calculated beam waists versus distance from focusing mirror

4 Thermal analysis and cooling scheme

For 42 GHz system which operates in pulsed mode mirrors are exposed to heat flux for short duration (500 ms). Hence no active cooling is required and heat can be dissipated in the thermal mass of the mirror material. However, it should be taken care that the temperature of the material does not reach near the melting point. So, transient thermal analysis is carried out for the mirrors. The maximum temperatures for 500 ms pulse are 301K and 303K for focusing and reflecting mirrors respectively. The temperature variation along with time for reflecting mirror is shown in the figure 6. However for 82 GHz system which can operate for 1000s, active cooling is mandatory. So, water flowing in the cavity absorbs the incident heat. The steady state thermal analysis is carried out for cavity like geometry for finding the distribution of temperatures on the mirror body in steady state.

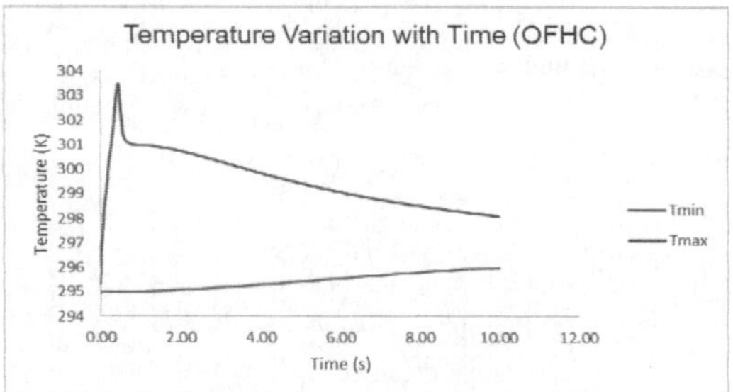

Fig 6. Variation of maximum and minimum temperature (K) of reflecting mirror with time (s)

The maximum steady state surface temperatures of focusing and reflecting mirrors with active cooling at flow rate of 6.8 liter per minute are 315K and 323K respectively. The contours for focusing mirror are shown in the figure 7.

Fig 7. Contour of steady state surface temperature (K) of focusing mirror

5 Structural analysis

The main loads acting on the mirrors are internal fluid pressure, dead weight of the mirror, load due to thermal expansion and electromagnetic forces during plasma disruption. Among these, internal fluid pressure and dead weight are significant and they are applied simultaneously in static structural analysis. The maximum von mises stress is around 10 MPa and hence well below the fatigue strength limit of OFHC. The stress contour for focusing mirror of 82 GHz system is shown in figure 8.

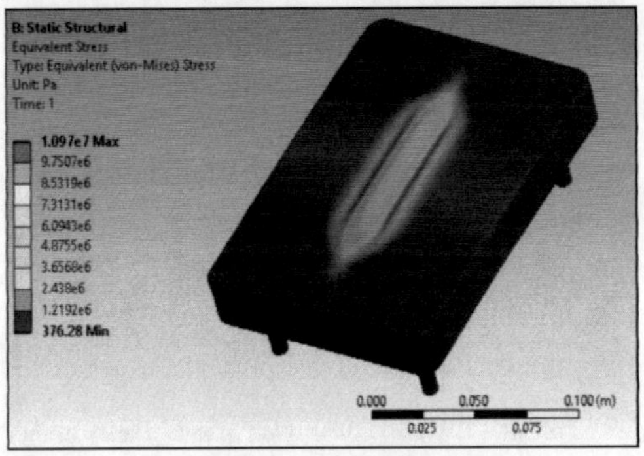

Fig 8. Von Mises stress (Pa) of focusing mirror

6 Conclusion

As per this design, the launcher consists of two focusing and two reflecting mirrors. The beam radii for 82 GHz and 42 GHz system at the resonance layer of plasma are 20 mm and 25 mm respectively. The thermal aspects of the launcher are studied and the maximum temperatures of focusing and reflecting mirrors with active cooling are 315K and 323K respectively for 82 GHz system. Since 42 GHz system is pulsed system, heat load is not critical for 500 ms operation. Hence design is able to withstand the thermal loads of SST-1 and it is safe. Similarly, to ensure the structural integrity the structural analysis is carried out considering the significant forces and design is found safe.

Further study is being carried out on the dynamic aspects of the system and optimization of the various parameters to reduce the response time of the mirror.

References

1. D. Bora, K. Sathyanarayana, B.K.Shukla et. al., Nucl. Fusion **46**, 3 (2006) pS72 - S84
2. D. Bora, K. Sathyanarayana, B. K. Shukla, P. Chattopadhyay, Y S S Srinivas, et al., Journal of Physics **25**, 1 (2005) p96-102
3. B. K. Shukla, P. Patel, J. Patel, R. Babu et. al., IEEE trans. on plasma sci., **vol. 43**, no. 1, (2015) p 485-488
4. B. K. Shukla, K. Sathyanarayana, P. Biswas et. al., Fusion Sci. and Tech., **v. 45(4)**, (2004) p. 549-55

EPJ Web of Conferences **147**, 04008 (2017) DOI: 10.1051/epjconf/201714704008
EC-19

Proposal of Mode Selection Criterian for Advanced KDEMO Gyrotron

Ashwini **Sawant**[1], Mun Seok **Choe**[2], and EunMi **Choi**[2,a]

[1] *Department of Electrical and Computer Engineering, UNIST, Ulsan, Republic of Korea.*
[2] *School of Natural Science, UNIST, Ulsan, Republic of Korea.*

Abstract. We proposed the design of a 255 GHz advanced gyrotron for Korea demonstration power plant (KDEMO) operation. Mode selection criterian for this DEMO gyrotron (output power > 1 MW) is discussed and a conventional cavity is designed for selected $TE_{55,13}$ mode. The designed cavity also excites the $TE_{37,9}$ mode at 170 GHz frequency. Beam wave interaction at both the frequencies are also analysed in our study.

1 Introduction

Gyrotron oscillators delievering MW's of power at millimeter wave regime are considered as a RF source for Electron cyclotron resonance heating (ECRH) and current drive (CD) in nuclear fusion reactors. DEMO [1] which is a futuristic ambitious step in thermonuclear fusion reactor field beyond ITER requires an advanced gyrotron for ECRH and ECCD [2]. The gyrotron is required to deliver MW of power along with the frequency tuning capability [3]. Conventional cavity gyrotron and coaxial cavity gyrotron are under investigation for 255 GHz operation. Here we have presented the mode selection criteria and initial design for a conventional type gyrotron cavity.

2 KDEMO Gyrotron Design

2.1 Mode Selection

To deliver high power of MW order at high frequency, a higher order mode must be selected. Apart from the conventional technique of mode selection, frequency tuning and output power criteria also play a major role in selection of the mode. Multi-frequency operation is only possible when all the operating frequencies are supported by the output window and the quasi-optical launcher. Output window is supported by multiple frequencies when these frequencies are its harmonics. In our gyrotron design, 170 GHz and 255 GHz are chosen as operating frequencies as these two frequencies are harmonics of 28.33 GHz and support ITER and DEMO operation respectively. For a feasible operation of quasi-optical launcher the selected modes must have the same caustic radius. $TE_{55,13}$ and $TE_{37,9}$ are the two modes that have almost same caustic radius and resonate at 255 GHz and 171.7 GHz respectively inside a conventional cavity of 20.78 mm radius. Fig. 1 shows the density of the mode spectrum around 255 GHz and relative couplings of the modes at optimum beam radius of 10.68 mm.

[a] e-mail: emchoi@unist.ac.kr

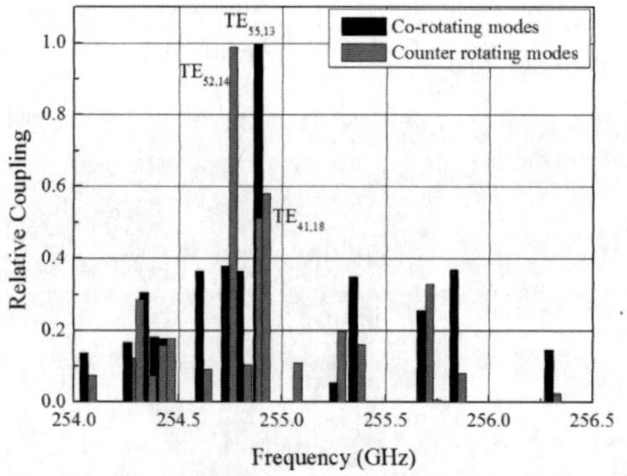

Figure 1. Relative coupling of the $TE_{55,13}$ and its neighboring modes for beam radius of 10.68 mm.

2.2 Cavity Design

Interaction of the beam and field takes place inside an open ended cavity in gyrotron. conventional cavity [4] and coaxial cavity [5] are the two popular type of cavities inherited for gyrotron design. In case of High Power Gyrotron ex. ITER and DEMO Gyrotron, which generally works on higher order mode to generate the MW's of power, coaxial type cavity has an advantage of reduced mode competition over conventional type of cavity. Gyrotron with coaxial type cavity can produce comparatively higher power but at the cost of more sophisticated design. We are analyzing the both type of cavities for 255 GHz Gyrotron but here we will just focus on conventional type of gyrotron cavity. A conventional gyrotron cavity is designed with an in house developed code "UNIST gyrotron design Tool (UGDT)" as shown in Fig. 2 which excites $TE_{55,13}$ mode at 254.99 GHz with a quality factor of 1186. We also performed the cold test of the cav-

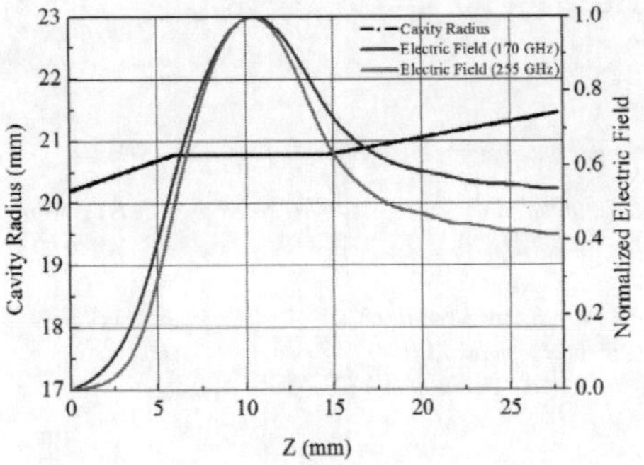

Figure 2. Normalized electric field along the cavity length.

ity at around 170 GHz and we found that $TE_{37,9}$ mode also resonates inside our designed cavity at 171.7 GHz but with a relatively smaller quality value of 560 as shown in Fig. 2. The geometrical parameters of this cavity are tabulated in Table 1.The beam wave interaction analysis is also carried out using time independent self consistent UGDT code. The beam wave interaction analysis shows maximum power of 1.12 MW at 10.16 T magnetic field for $TE_{55,13}$ as shown in Fig. 3 for a high quality electron beam as tabulated in Table 2. It also predicts a maximum power of 1.5 MW at 6.35 magnetic field for $TE_{37,9}$ mode as shown in Fig. 3.

Table 1. Cavity Geometry

Design Parameter	Value
Input Taper Radius (R_1)	20.2 mm
Cavity Radius (R_0)	20.78 mm
Output Taper Radius (R_3)	21.5 mm
Input Taper Length (L_1)	6 mm
Mid-section Length (L_2)	8.7 mm
Output Taper Length (L_3)	13 mm

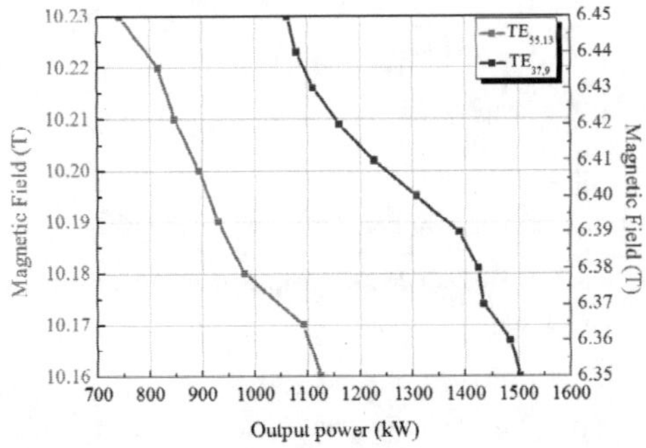

Figure 3. Magnetic Field Vs Output Power for both modes.

Table 2. Gyrotron Design Parameters

Design Parameter	Value
Beam Voltage (U_b)	80 kV
Beam Current (I_b)	40 A
Alpha (α)	1.3
Magnetic Field (B_0)	10.2 T
Beam Radius (R_b)	10.68 mm

3 Conclusion

A rigorous analysis of mode selection for KDEMO gyrotron is carried out to generate MW of power at multi-frequencies by using a UNIST gyrotron design Tool (UGDT). The cavity design and its interaction with the beam is also analyzed using UGDT. The design of the quasi-optical launcher and RF window is still under investigation.

References

[1] K. Sakamoto, K. Takahashi, A. Kasugai, R. Minami, N. Kobayashi, S. Nishio, M. Sato, K. Tobita, Fusion Engineering and Design **81**, 1263 (2006)

[2] J. Jelonnek, K. Avramidis, J. Franck, G. Gantenbein, K. Hesch, J. Jin, P. Kalaria, A. Malygin, I.G. Pagonakis, T. Rzesnicki et al., in *Infrared, Millimeter, and Terahertz waves (IRMMW-THz), 2014 39th International Conference on* (2014), pp. 1–3

[3] P.C. Kalaria, K.A. Avramidis, J. Franck, S. Illy, I.G. Pagonakis, M. Thumm, J. Jelonnek, in *Vacuum Electronics Conference (IVEC), 2015 IEEE International* (2015), pp. 1–2.

[4] A. Sawant, S.G. Kim, M.C. Lin, J.H. Kim, Y. Hong, J. So, E. Choi, IEEE Transactions on Plasma Science **42**, 3989 (2014)

[5] B. Piosczyk, A. Arnold, G. Dammertz, O. Dumbrajs, M. Kuntze, M.K. Thumm, IEEE Transactions on Plasma Science **30**, 819 (2002)

The Experimental Verification of Gaussian Beam Coupling for ECH Transmission Line at 400 GHz

Mun Seok Choe[1,a], Ashwini Sawant[2] and EumMi Choi[1,2]

[1]*Physics and Applied Mathematics, 44919 Ulsan National Institute of Science and Technology (UNIST), Republic of Korea*

[2]*Department of Electrical and Computer Engineering, 44919 Ulsan National Institute of Science and Technology (UNIST), Republic of Korea*

Abstract. *We design a quasi-optical transmission line system for a 400 GHz gyrotron beam. The 400GHz Gaussian beam is injected to a corrugated waveguide bounced from a quasi-optical mirror. From detailed 2D field patterns of the output beam emitted from the corrugated waveguide, we analyze the mode contents and the source of non-ideal beam expansion*

1 INTRODUCTION

In sub-terahertz (THz) range, development of high power transmission lines is faced with problems such as manufacturing tolerance and coupling efficiency. For preparation of KDEMO, a proceeding research is required to transfer ECH beam to tokamak with low loss at sub-THz frequency. A typical case is a Gaussian beam coupling from a gyrotron ouput to a corrugated waveguide. A HE11 mode in the corrugated waveguide can reduce an Ohmic loss, and the coupling of Gaussian beam coming from gyrotron to the HE11 mode is important [1]. In this research, we show the gyrotron output Gaussian beam coupling through a quasi-optical transmission line and measurement of 2D field patterns coming out the corrugated waveguide.

2 THE COUPLING OF GAUSSIAN BEAM AND HE11 MODE

The quasi-optical transmission line is fabricated as shown figure 1. The system consists of a quasi-optical ellipsoidal mirror, a corrugated waveguide, and a 400GHz ultra Gaussian beam antenna from Thomas Keating Ltd. Based on the beam waist location of the Gaussian beam antenna, the quasi-optical mirror and corrugated waveguide is designed by Surf3D simulation code [2].

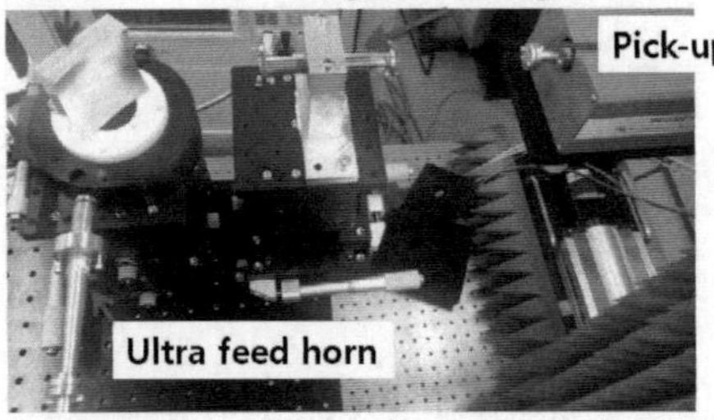

Figure 1. Experimental setup of the quasi-optical transmission line at 400 GHz

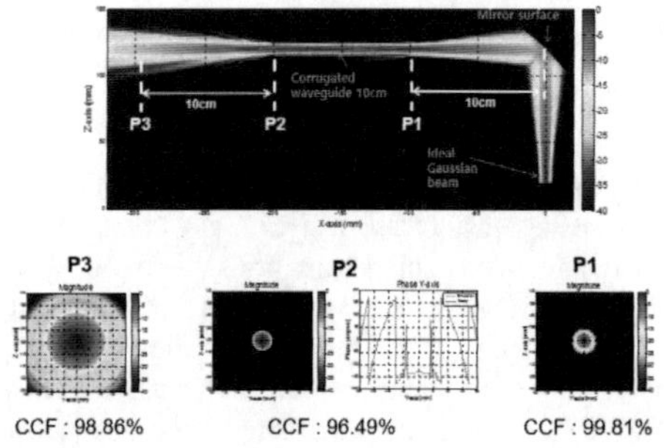

Figure 2. Surf3D simulation results (P1 : input of corrugated waveguide, P2 : output of corrugated waveguide, P3: 10cm distance from output)

Figure 2 shows results of Surf3D simulation. The beam waist from the aperture of the Gaussian beam antenna is 2.73 mm and the output beam waist reflected by a mirror is also 2.73 mm for coupling with HE11 mode in the corrugated waveguide. Compared to theoretical results, gaussian beam is coupled with corrugated waveguide in high efficiency. In experimental setup, a precise alignment is dependent on accuracy of input beam. Thus, we measured the field pattern from ultra feed horn antenna by using a Vector Network Analyzer (N5247A) as the open waveguide is automatically adjusted by a motion controller. As illustrated in Figure 3, the gaussian beam propagation is followed to theroretical graph. Specially, the field pattern in mirror reflectance region (10 cm distance from the ultra feed horn) is well matched.

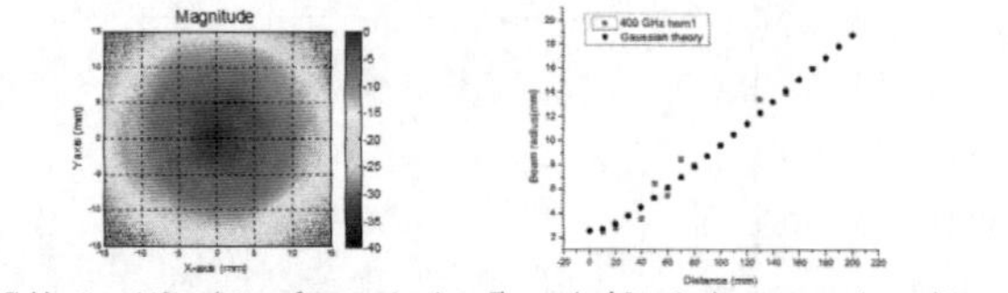

Figure 3. Gaussian beam propagation from ultra-feed horn antenna

Figure 4 shows field pattern of beam waist for coupling with the corrugated waveguide. The magnitude of electric

[a] Corresponding author: emchoi@unist.ac.kr

field is a little bit distorted but the phase is almost flat. Based on the results, corrugated waveguide is set up to beam waist position.

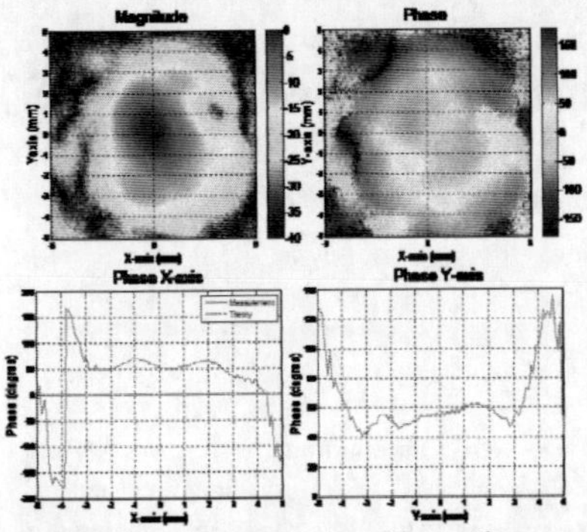

Figure 4. The beam patterns at 10 cm from an ultra-feed horn antenna (Beam waist position, an entrance of corrugated waveguide)

Figure 5 presents the propagated beam pattern at 10 cm distance from the aperture of corrugated waveguide. When compared with simulation results, the experiment data shows that the measured output beam agrees well with the Gaussian distribution although some distortion of the field pattern has been observed. A cross correlation function (CCF) is analyzed to be around 89 %. Considering a few μW levels and the calibration error ($\pm$0.1dB), some discrepancy between the measured data and ideal Gaussian case can be analyzed in sub-terahertz regions.

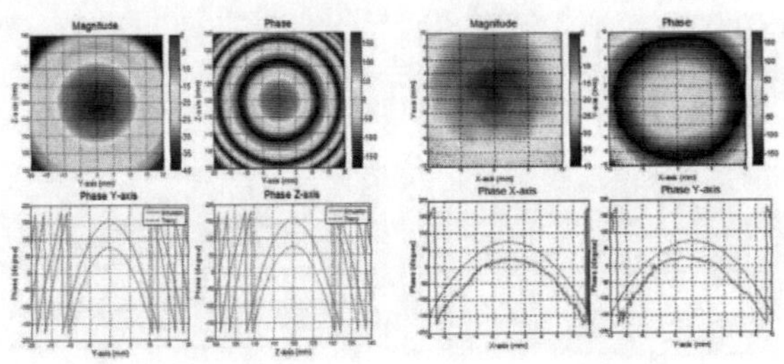

(a) (b)

Figure 5. The beam patterns at 10 cm from an aperture of the corrugated waveguide (a) Surf3D simulation results (b) Experimental results

3 CONCLUSION

The study of the coupling with Gaussian beam and HE11 mode is performed in sub-terahertz transmission line. The experimental data have been compared with simulation results and confirmed a reasonable agreement. Futhermore, we can identify THz field measurement system using open ended waveguide and VNA. Through the phase correcting mirror, the system will be upgrade for high efficiency. Finally, we will proved the possibility of THz ECH transmission line in high power. These results may help to develop advanced technology for KDEMO project.

References

1. J Kowalski, Elizabeth J. et al. "Linearly Polarized Modes of a Corrugated Metallic Waveguide." IEEE Transactions on Microwave Theory and Techniques, Vol.**58**, NO. 11, pp 2772–2780, 2010.
2. J. M. Neilson, and R. Bunger, "Surface integral equation analysis of quasi-optical launcher," IEEE Transactions on Plasma Science, Vol. **30**, pp 794, 2002.
3. Mun Seok Choe et al. "A Comprehensive Analysis of a TE11 to HE11 Mode Converter for an Oversized F-band Corrugated Waveguide", Journal of Electromagnetic Waves and Applications, Vol 27, Issue 17, pp 2221-2238, 2013.

EPJ Web of Conferences **147**, 04010 (2017) DOI: 10.1051/epjconf/201714704010
EC-19

Instrumentation and control system architecture of ECRH SST1

Harshida Patel, Jatin Patel, Dharmesh purohit, B K Shukla, Rajan Babu, Hardik Mistry and ECRH team
Institute for Plasma Research
Bhat, Gandhinagar-382428
Email: harshita@ipr.res.in

Abstract. The Electron Cyclotron Resonance Heating (ECRH) system is an important heating system for the reliable start-up of tokamak. The 42GHz and 82.6GHz Gyrotron based ECRH systems are used in tokomaks SST-1 and Aditya to carry out ECRH related experiments. The Gyrotrons are high power microwave tubes used as a source for ECRH systems. The Gyrotrons need to be handled with optimum care right from the installation to its Full parameter control operation. The Gyrotrons are associated with the subsystems like: High voltage power supplies (Beam voltage and anode voltage), dedicated crowbar system, magnet, filament and ion pump power supplies and cooling system. The other subsystems are transmission line, launcher and dummy load. A dedicated VME based data acquisition & control (DAC) system is developed to operate and control the Gyrotron and its associated sub system. For the safe operation of Gyrotron, two level interlocks with fail-safe logic are developed. Slow signals that are operated in scale of millisecond range are programmed through software and hardware interlock in scale of microsecond range are designed and developed indigenously. Water-cooling and the associated interlock are monitored and control by data logger with independent human machine interface.

1 Introduction

The electron cyclotron resonance heating (ECRH) is used for plasma start-up to MHD control in various tokamaks [1-4]. The 82.6GHz/200KW and 42GHz/500kW ECRH systems are used in tokamak SST-1 [5-6] to carry out experiments related ECRH assisted breakdown and start-up at fundamental and second harmonic. The operating toroidal magnetic field of SST-1 is 0.75T, 1.5T and 3.0T, however in the beginning phase of tokamak, it is operated at 0.75T and 1.5T magnetic fields. The 42GHz/200KW system is used for starting phase of operation of SST1 [7]. Both ECRH systems 42GHz and 82.6GHz ECRH system consists of Gyrotron, transmission line and a mirror based launcher and other auxiliary power supplies. The 82.6GHz Gyrotron and 42GHz Gyrotron are capable to deliver 200KW/CW and500kW/500ms at 50kV beam voltage respectively.

The system is connected to tokamak using corrugated waveguide based transmission line. The transmission consists of matching optic unit, bends, polarizer, DC break and mirror based launcher. The layout of 42GHz ECRH system and actual system in SST-1 tokamak hall are shown in figure1 and 2 respectively.

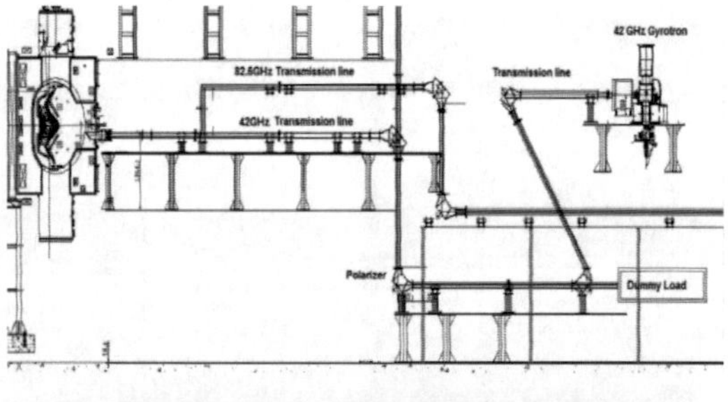

Figure1. 82.6GHz and 42GHz ECRH system on SST1.

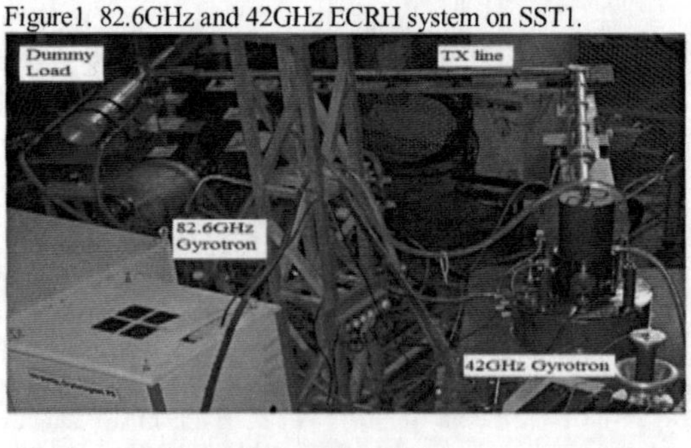

Figure2. ECRH system in SST1 Tokamak

[*] Corresponding author: harshita@ipr.res.in

DOI: 10.1051/epjconf/201714704010

2 System Requirements

The Gyrotron based ECRH system is associated with many auxiliary power supplies like filaments, high voltage cathode power supply, magnet power supply, anode power supply for generation and control of microwave power. Because of high power generation in Gyrotron microwave tube, sufficient water cooling is provided in component of gyrotron like collecter, body, cavity for heat deposition .The dedicated data acquisition and control (DAC) system is developed to monitor the entire system and perform the automated system tasks [8]. The DAC monitors parameter of axillaries power supplies and controls the according to the system requirements. A dedicated datalogger is used to monitor all cooling parameter of system.The Control system has two parts conditioning instrumentation and interlocks which are discussed in next sections.

3 Conditioning Instruments

The Gyrotron based ECRH system consists auxiliary power supplies, microwave components and cooling system. All the different systems have different control parameters and Different operating range. A single control platform is required to operate whole system automated. The conditioning electronics is developed to convert all parameters to common control range for central control system of DAC. The Conditioning electronics is the common interface between auxiliary system and central control system .The ECRH DAC has VME based central control system, which controls the subsystem.

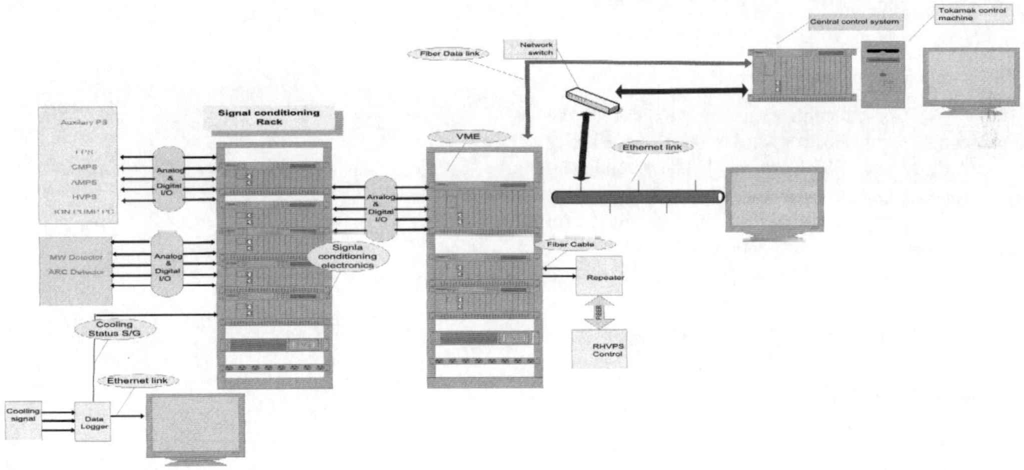

Figure 3. Conditioning instruments of ECRH system.

The Common signal flow diagram of conditioning system is shown in figure 3.In the figure different auxiliary power supplies and cooling control parameters are connected to conditioning electronics.

The Local DAC system of ECRH is interfaced with central control system of tokamak SST1 for various plasma experiments. The optical based Ethernet protocol is used to interface between ECRH and Tokamak control unit. The VME central unit and data logger are connected with Ethernet link with GUI (graphical user interface).

The ECRH DAC control system handles around 80 signals for monitoring and control, which include cooling parameter and interlock signals. The safe operation of Gyrotron requires various Input and Output signals (Analog & Digital). In order to operate the system in High voltage environment, Fiber based input and output interface is used to avoid the transfer of noise with the signal. The VME operate sequential step to interface with local unit of high voltage power supply. The indigenously developed instrumentation and control fulfil the requirements for ECRH operation.

4 Interlock systems

The interlock system operation depends on the hierarchy and critical requirement with different time varies from microsecond to millisecond range. The slow scale interlocks action taken care by VME control unit. This uses software coding and called as soft interlocks. The slow scale interlock associated cooling is operated with Data logger controller unit. The operational window for cooling parameter (Flow & temperature) are decided for safe operation and interfaced with control unit. Topology of interlock unit is shown in figure4.

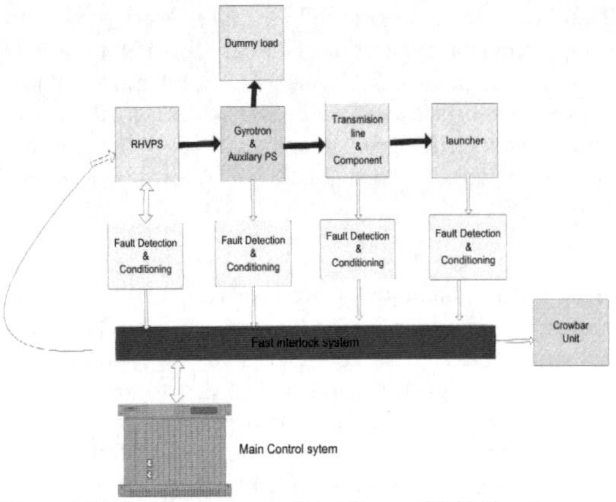

Figure4.fast interlocks processing scheme of ECRH system

The Gyrotron is delicate device and its safe operation demands dedicated fast interlock for the protection of Gyrotron. There are several fast interlock are used for safety of Gyrotron. These are Arc, beam over current, di/dt etc. In case of fast interlock operation, the main high voltage power supply need to be removed within 10μs scale. All ARC detection unit and main power supply parameter like di/dt and beam over current are directly interfaced with dedicated hardwired control system. The fast interlock unit works as an independent control system provides a single trigger to main HV power supply to protect Gyrotron against any damage. Since the time of operation is critical (<10μs) so sensors are directly connected to interlock unit. The optical fiber based communication is used to Interface with interlock and main HV power supply to minimize the delay for reliable and safe operation.

Summary

The VME based control system along with the required conditioning electronics has been successfully implemented in ECRH SST1 system. This is a stable and reliable system operates continuously with Gyrotron. This system is being upgraded with PXI based control system [9]. The fast interlock action will be based on FPGA based processor where the signals would record the sequence of event for post event analysis.

Acknowledgement

The authors would like to thank ECRH team for extended cooperation during entire Development and implementation phase of control system. The author would like to thank all the members of Main high voltage team for their support for the implementation of DAC system.

References

1. B. Lloyd et al., 'low voltage ohmic and Electron Cyclotron heating assisted startup in DIII-D' *Nuclear Fusion,* Vol.**31**, No: 11, 2031-2053, 1991.
2. F. Leuterer et al. "Recent ECRH results in ASDEX Upgrade" *Nuclear Fusion,* Vol. **43** (2003), 1329-1342
3. M. Rome, V. Erckmann, U. Gasparin and N. Karulin, "Electron cyclotron resonance heating and current drive in the W7-X stellarator " *Plasma Phys. Control. Fusion* **40** (1998) 511-530.
4. R. Wilhelm et al., *Plasma Physics and Controlled Fusion,* **26** (1984) pp 259-261
5. D. Bora et al. "Test and commissioning of 82.6GHz ECRH system on ST1" *Journal of physics*, Vol **25** (2005) page(s)96-102
6. B. K. Shukla et al." 42HZ/500KW Electron cyclotron Resonance Heating on Tokamak SST1", *IEEE TPS*, volume:**43**, Issue :1 ,January (2015)
7. B. K. Shukla et.al. "42-GHz 0.5-MW ECRH System for Tokamaks SST-1 and Aditya", *IEEE Trans. on Plasma Science* Volume: **40** , Issue: 4 Page(s): 1234 – 1238
8. K. Rajnish et al, "VME based real time Data Acquisition and control for 82.6 GHz Gyrotron based ECRH system on SST1 tokomak " *20th national symposium on plasma science society of India, PLASMA -2005*,Cochin –India. FT -5.27
9. J. Patel et. al., "Commissioning of PXI based Data acquisition and control system for ECRH", *EC19*, paper tech-P-7

AUTHOR INDEX

AUTHOR INDEX